CALF REARING

◆ Bill Thickett ◆ Dan Mitchell ◆
◆ Bryan Hallows ◆

The Crowood Press

First published in 1986 by
The Farming Press, Ipswich

Revised edition 1988

This edition published in 2003 by
The Crowood Press Ltd
Ramsbury, Marlborough
Wiltshire SN8 2HR

www.crowood.com

© Farming Press Ltd 1986 and 1988

British Library Cataloguing-in-Publication Data
A catalogue record for this book is available from the British Library.

ISBN 1 86126 643 X

Typeset by Galleon Photosetting, Ipswich

Printed and bound in Great Britain by Biddles Ltd

CONTENTS

v

Acknowledgements

I AM INDEBTED to my colleague Nick Cuthbert for his invaluable help in reading the manuscript and making valuable suggestions. Also to Jayne Mosford for her skill in deciphering my handwriting to produce a fair and readable copy.

BILL THICKETT

Illustration credits

Plates: 1–12   *Dairy Farmer* magazine.
        13, 16, 17, 19, 20   Dan Mitchell.
        14, 15, 18, 21, 22   NAC Calf and Beef Unit, Stoneleigh.
        23–29   *A Veterinary Book for Dairy Farmers*, R. W. Blowey,
                Farming Press.

Jacket: calf drawn by Louise Dunn.
        design by Hannah Berridge.

Introduction

THIS BOOK seeks to provide a practical guide on rearing calves to twelve weeks of age; it is for calf rearers, prospective rearers and students of calf rearing. The experience of the three authors in their own fields of cattle development work involving large numbers of calves, advisory work on farm buildings, and veterinary practice in an intensive calf-rearing area, is given in a form which it is hoped will be easy to read and digest. The aim is not an exhaustive reference book since these are already available, but is to reflect proven modern practice as experienced in a wide variety of situations.

The rearing of calves is following the path of intensification already taken by the pig and poultry industries. There are several reasons for this. The dairy herds, from which the majority of rearing calves come, are themselves very much bigger than they were even ten years ago. This has progressively created problems for some herdsmen through pressure of extra work at busy calving times. Unfortunately they are often given adequate new facilities for the expanded number of cows, but asked to manage with existing or limited expansion of accommodation for calves.

Not only does this have to house the larger number of heifer replacements required for the expanded dairy herd, but it is often expected to accommodate surplus calves which are going to be sold for the beef market at home or for the export trade, largely for veal. These calves for sale need a very good start in life, but in cases where the housing is not designed for mechanised cleaning out, or where a periodic rest is not possible, there is a build-up of infection throughout the winter which can become so bad as greatly to increase illness and mortality.

It is hoped that the chapter on housing will provide a guide to those who wish to convert old buildings or provide new accommodation. Mistakes in design lead to poor performance and health and can be very costly to rectify. It is better to avoid them if possible by learning from the experience of others summarised here.

With the imposition of milk quotas the bull calf for sale assumes an

1

even greater importance as a source of income. Together with the new code of practice for the marketing of calves we would expect an improvement in their presentation with advantage both to the buyer and seller.

Some dairymen who adjusted to their milk quota by reducing the number of cows have opted to replace them with a small beef enterprise, and we hope that they will find practical help in this book to make it successful.

In addition to those rearing calves on dairy farms, there are a growing number of specialist calf rearers who rear calves mainly on contract. Their operation is to buy in calves at a vulnerable early age and turn them into level lots of reared calves which the beef finisher can purchase at twelve weeks of age, knowing that their calfhood problems are behind them.

These specialised rearers are themselves under pressure to do the job well but require a quick turnover with a minimum of labour. They cannot survive unless mortality is low and results are good. Accordingly they have become very professional in an intensive situation. They usually rear calves by bucket feeding and are very cost conscious.

Because their calves go direct to a beef finisher, a reputation for subsequent good performance is a guarantee for another order next year and one which is well worth cultivating. This reputation is partly based on selection of the right type of calves but also on giving them a flying start which seems to establish them for a good second phase.

Part of the response on the dairy farm to intensification and particularly the need to save labour has been to adopt an ad lib system of rearing. Some five years ago the cold acidified system became popular, but this has now been partly replaced by the present trend towards automatic machine feeding, especially on the larger farms. The pros and cons of these systems with regard to performance and economics are discussed.

There have been significant changes also in the formulation of milk replacer diets and dry concentrate feeds in the last five years and these are still continuing. The rapid and substantial increase in the price of skim milk powder has led to a search for milk protein substitutes and the promotion of 'zero' skim milk powder diets.

However the majority of milk replacer diets are still based on 60 per cent or more of skim milk powder, often acidified and imported. Other milk replacers imported from the Continent, often associated with the provision of 60 kg liveweight calves for export, are veal-type diets. This has led to a profusion and confusion of brands, so that it is not easy to discover their reliability. The calf itself, whether destined for the dairy herd, veal or beef, has more than kept its value in the inflationary spiral

and justifies the correct feeding, housing and health-promoting pro-
grammes to ensure its successful rearing. It is too valuable not to be
looked after as well as possible. Sample costings shown in the
Appendices highlight the relative importance of feed costs, other
variable costs and overhead costs in the overall picture of profitability.
The headings are given as a guide.

We recommend good record-keeping to the rearer and the construction of one's own feed and other costs and profit into an acceptable
selling price. This can be done either as an individual or as a member of
a group, where bulk purchases can lead to economy of feed inputs. The
actual prices given are soon outdated of course, but they show the
relative importance to that all-important 'bottom line'. We wish all our
readers 'Good Calf Rearing'.

Chapter 1

SYSTEMS OF REARING

GENERAL CONSIDERATIONS

Measuring Efficiency

ALL SYSTEMS of calf rearing demand high standards of performance achieved as economically as possible. It is fortunate that in the period up to twelve weeks, the age at which the calf is considered to be 'reared', good rates of growth are synonymous with economic performance.

From twelve weeks the fully ruminant young animal indoors can utilise a variety of farm feeds. The most typical would be hay or silage supplemented by concentrates.

The actual cost and efficiency of growth up to twelve weeks are not difficult to establish. However, even during this early period the efficiency of feed utilisation declines as the calf changes from functioning as a simple-stomached animal, digesting whole milk or milk replacers very efficiently, to a ruminant animal involving all four stomachs. When fully ruminant the calf's digestive process involves bacterial fermentation in the rumen, with further digestion and absorption taking place along the alimentary tract.

Efficiency of feed utilisation can be expressed as Feed Conversion Ratio (FCR) in terms of kg of feed required per kg liveweight gain (LWG). When giving feeds of varying dry matter (DM) the ratio is stated as kg feed DM per kg LWG and this is an accurate measure of efficiency. Similarly the appetite of a ruminant animal is quoted in dry matter terms during ration formulation.

More precisely still, efficiency can be expressed as units of energy in megajoules* (MJ) per kg of liveweight gain. The metabolisable energy (ME) system, now used in ruminant nutrition, uses the joule as the basic unit, replacing the old starch equivalent system and the calorie.

The rearing period from twelve weeks onwards is more difficult to monitor accurately and efficiency of feed use depends very much on the provision in summer of high-quality grazing and in winter of fodder

*(1 calorie = 4.184 joules; 1 megajoule = 1 million joules).

Table 1.1 Changing efficiency of the young calf

Period	Feed conversion ratio	MJ for maintenance and 1 kg LWG
0–5 weeks	1.9:1	25
5–12 weeks	3.0:1	31

correctly supplemented with concentrates. The feed conversion ratio will continue to fall, however well-managed the animal, until it reaches maturity.

Target Liveweight Gains

There is now adequate nutritional data to plan the required rate of growth for dairy heifers to calve at the desired weight and age, and similarly for beef animals to reach marketable weights, correctly finished, when prices are high.

Many calf rearers, whether rearing for beef or dairy replacements, have now adopted a system aimed at giving the calf a very good start in life. This has proved to be economic, and has enabled targets to be met for calving Friesian heifers at two years of age weighing 500 kg. Similarly beef animals can reach 450–500 kg at 16–18 months on a semi-intensive system, or 420 kg on an intensive system in less than one year.

This good start could be defined as at least doubling the calf's birthweight in twelve weeks.

For example: A Friesian heifer calf born weighing 40 kg will weigh 82 kg at twelve weeks where it has put on 0.5 kg/day liveweight gain which is well within its potential.

A Friesian bull calf weighing 45 kg at birth or purchase will weigh 108 kg at twelve weeks which requires 0.75 kg per day liveweight gain, which again is quite achievable with good management.

Growth Potential

The potential for growth is not uniform for all cattle, uncastrated bull calves having about a 12 per cent advantage over heifers. Additionally, the birthweight of the calf has an effect on subsequent growth rate with the heavier calf able to put on weight faster than the lighter one when both are given adequate nutrition. Each difference of 1 kg in birth weight will expand to about 1.5 kg difference by twelve weeks, given

the same management. It will of course eat slightly more food to do this.

Of much greater magnitude are the effects of feeding, health and management throughout, and over which the farmer has a good measure of control.

The Main Systems of Rearing

There are many different methods of rearing adopted by farmers. I suspect that some of these are decided by tradition and by available housing and equipment rather than by a conscious evaluation. Besides the target liveweight gains already given the systems have some general features in common.

For roughly the first five weeks the principal nutrition for the calves comes in the form of liquid feed, either whole milk, milk replacer or acid-milk replacer. Water and dry food are made constantly available. The calves are then weaned off the liquid feed and on to the dry food which will see them through to twelve weeks.

There is usually a lively interest and discussion about new ideas and developments at farmers' and rearers' meetings. The introduction of acidified diets for cold ad-lib feeding is a good example of a new system which certainly generated a lot of controversy and discussion. It set new standards of performance, and generally focused a good deal of attention on calf rearing as a whole, especially amongst specialist calf rearers. For them a high degree of expertise in an efficiently run unit is essential to their livelihood.

The main systems which will be considered in this chapter and throughout the book are as follows:
1. Bucket rearing using whole milk or milk replacers given:
   (a) once a day
   (b) twice a day;
2. Ad-lib feeding:
   (a) cold, using acid milk replacers
   (b) warm (machine feeding).

Colostrum

Whatever the system of rearing, it is vital for disease resistance that the calf receives adequate colostrum from its dam. It is best if the calf obtains this by sucking during the first few hours of life. The calf may need help in finding the teat. If it is found to be impossible for the calf to suck, perhaps because of a very low udder, or because a first-calving heifer is fidgety, then some colostrum should be drawn and given to the calf by bottle or by oesophageal tube.

By the second day, when it is likely that the calf is removed from its dam, an easy way to continue with a good intake of colostrum is to use a communal container fitted with self-closing teats. A group of calves can then be given warm colostrum/milk twice a day and allowed to help themselves.

An indication of the daily intake came from one batch of seventy-six Friesian calves of both sexes with an average birthweight of 43 kg (95 lbs) reared at Barhill. In an average of 4½ days on the ad-lib container, each calf took 7.8 litres (1¾ gallons) per day.

BUCKET FEEDING

This is by far the commonest system on all types of farm. It owes its popularity to the fact that it is a simple system to operate and that calves can be taught to drink from a bucket very soon after birth, In fact the sooner after twenty-four hours the change is made from sucking the dam to bucket feeding, the easier it is. A day or two on a teat feeder as described above will not complicate matters; the important move is to take away the maternal influence.

The dairy farmer has good control over this early period, changing the calf from its dam to hand-fed colostrum, then to whole milk and finally to milk replacer within the first week of life.

Calves can be extremely lively and difficult to handle. The stockman has more control of the calf where a single pen is available because he can back it into a corner between his knees. He can then guide the calf's nose into the bucket by using his finger as a substitute teat.

The amount of milk given to the calf in the first few days will depend on how quickly it learns to drink. Spillages often make it difficult to know exactly how much is consumed, but the rearer should aim to get the calf to drink at least 3 litres per day (⅔ gallon) over the first few days.

Use of Whole Milk

At this stage of rearing the dairy farmer with access to whole milk has to decide whether to use this or switch to a milk replacer. A recent survey has shown that 14 per cent of dairy calves are reared on whole milk.

It is perhaps surprising that there should be any doubt in opting to use milk replacer because of their relative costs. For the last thirty years whole milk and milk replacers have kept a fairly constant cost ratio of 2:1. In 1950 milk was 2 shillings per gallon whereas milk replacer was 1 shilling per gallon when mixed at 1 lb/gallon (10 per cent). In 1988 with milk at 16p/litre, a top-quality milk replacer when mixed at 100 g/litre (also 10 per cent) cost 8.5p/litre. However, by 1995 the cost of whole milk at 24p/litre has gone down to only 1½ times the cost of milk replacer at 15.5p/litre. Incidentally modern milk replacers are vastly more nutritious than their forerunners thirty years ago, when the early-weaning system first evolved.

Where only a small number of calves are reared, a dairy farmer can make out a case for the use of whole milk, overriding the less favourable economics. Performance on milk will be good in the early stages, and controlled trials have shown it to be slightly superior to milk replacers. Whole milk was, after all, designed for the job, and its digestibility in the calf's true stomach (abomasum) is extremely high at 94 per cent. It also contains a naturally occurring anti-bacterial system protecting calves against infection. These may also be present in milk replacers if they are manufactured to high standards of quality control during the drying process.

Dry feed intake is less on whole-milk feeding than on milk replacer feeding, and care is essential to ensure adequate concentrate intake at weaning, or there will be a check in performance.

Bull calves intended for immediate sale may often, especially when prices are high, be left suckling the cow for a week or ten days in order to present a 'good-backed' calf showing plenty of bloom in the market. On larger farms, particularly, this practice, which is not in the best interests of the dairy farmer, is under challenge from the machine feeding of milk replacer. This carries the calves on without the need to arrange for calves to suckle their dams which can go through the parlour instead, and earn more income.

Milk Replacers

The vast majority of rearers who are raising heifer replacements or beef calves now use milk replacers in various systems. Top-grade powders have a high digestibility—around 92 per cent and arc based on a minimum of 60 per cent skim milk powder inclusion. More details of these are given in Chapter 3.

There is some confusion about *mixing strengths*, and it should be noted that mixing strengths quoted refer to the quantity of powder within the amount of the reconstituted mix, i.e. for one calf 250 g of milk powder should be mixed and *made up to* two litres with warm water. If two litres of water are added to 250 g of milk powder, the final mix is 2.25 litres and the calf has to take in slightly more liquid for the same nutrient intake. The important factor is to make sure that the correct amount of milk replacer is measured, or preferably weighed out, for the number of calves being fed.

For the record, where two litres of water are added to 250 g of powder, the strength is one in nine or 11.1 per cent, instead of one in eight or 12.5 per cent where the weighed 250 g is made up to two litres.

Twice-a-day Bucket Feeding

This is the system practised by the majority of bucket feeders. The reasoning behind this is quite logical, namely that it is closer to the natural system of suckling several times a day than is less frequent feeding. It is allied closely to twice-a-day milking on dairy farms, and it provides a compulsory check twice a day on calf health.

After four days on cow's milk, a gradual change over a period of two days to milk replacer can be made. In week two of the calf's life it will normally be taking a fixed amount of two litres (3½ pints) per feed with a mixing rate of 125 g/litre (2½ oz per pint).

For simplicity, and to avoid errors, it is strongly recommended that the required quantity of powder for a group of calves is weighed out using a bucket and spring balance. This can then be placed in a calibrated container with some water already in and mixed, either mechanically or by using a hand whisk. More water is then added to give the correct final volume and temperature. It must be said that this is easier to practise where the metric system is in use rather than imperial units.

The alternative to weighing is to use a measure, often provided by manufacturers, where one measure of milk replacer powder is equal to a feed for each calf. In this case the measure must be regularly check-weighed, because milk replacer powders can vary considerably in bulk density and errors of up to 20 per cent can arise.

It is very important that a consistent feeding temperature is used. This should be around blood heat or slightly below, but not above, i.e. about 36 °C or 97 °F.

Once the correct amount of liquid feed has been prepared, correct quantities can be measured out in buckets for individual calves using a litre or pint measure with a handle. Small 5 litre (1 gallon) plastic buckets with handles are ideal for feeding and are easy to keep clean. Three in each hand can easily be carried and placed individually in front of calves. For larger numbers of calves, a mixer on wheels with an electrically-driven bottom agitator can be filled at the mixing point, and then wheeled round with a supply of buckets, reusing them round the house.

With 50–100 calves the time taken will be about half a minute per calf where one man uses this method which is also applicable for once-daily feeding. Some might argue that each calf should have a *clean*

Plate 1 The amount of liquid for each calf is being measured from a mobile mechanical mixer. Straw is on offer in the racks and water is available in front of each calf. The milk buckets are placed on top of the calf feed to encourage the calf to take dry food after the milk bucket is removed.

bucket and certainly this will help minimise the transfer of disease organisms, provided that the washing up is not neglected. Thorough cleaning is absolutely essential and requires, ideally, a washing-up tank equipped with hot and cold taps. A suitable dairy detergent steriliser powder should be used.

To be thoroughly controversial, this rule of cleanliness has been broken for the last eight years at Barhill. Each calf has a clean bucket for milk replacer at the start, which stays in front of each pen for five weeks, *without ever being washed up*!

The adoption of this scheme followed an outbreak of salmonellosis brought in by the calves. Since this organism is so infectious, it was thought preferable not to mix buckets for washing up, but rather to prevent cross-contamination by keeping each calf's bucket separate. So far the system has worked well and there has not been any suggestion of it causing stomach upsets. Indeed one might even say the small residues in the bucket could be growing a vast colony of streptococcal probiotics!

The provision of warm water for feeding is an important practical consideration. The quantity required will be three litres per calf on once-a-day feeding or four litres on twice-a-day feeding and is best provided from an insulated tank with a thermostatically controlled electric heating element. This can operate most economically by heating overnight on the Economy 7 electricity tariff, as cheap rate electricity costs less than half the price of the daytime rate.

The temperature of mains water varies from about 4°C in winter to 11°C in summer. Normal practice is to heat water to 70°C and mix it with water from the tap, roughly at a rate of 2:1 to produce the final mix at 40°C, judged by hand or by testing with a thermometer. Some milk powders will only mix well at much higher temperatures than blood heat, so here the mixing can be done with the hot supply, cooling afterwards with cold water.

The size of the hot-water tank should be matched to the maximum likely to be required, e.g. for three litres of mix, two litres of hot water are needed. For thirty calves this would therefore be sixty litres (thirteen gallons). A normal domestic cylinder three feet high by one foot three inches diameter is about 114 litres and would therefore comfortably provide for thirty calves and for washing up. The cost of heating this tank once a day to 85°C (to allow for tank losses) is about 83p on standard rate or 28p on night rate, i.e. 97p or 33p respectively per calf over a five-week rearing period (once-a-day feeding).

Returning to the actual feeding, the chapter on housing will describe the most common type of single pen equipped with a dry feed bucket and a water bucket, both in holders *outside* the pen. The liquid feed bucket can be dropped into the dry feed bucket for feeding and to eliminate the need to interchange buckets. When the empty liquid feed bucket is removed the calf will tend to put its head back into the dry feed bucket and this encourages early intake of dry feed. The keen rearer often finds time to place a little dry feed in the calf's mouth at this point in the early stages of rearing.

The bucket containing water should remain in front of the calf at all times. A few calves may continue their urge to drink by consuming water. If this becomes excessive some rearers remove the water for an hour or so after liquid feeding, but this is not vital. Data on water intakes are given in Chapter 3.

Having fed the calves, wash up the equipment before residual liquid has time to dry, and then observe the calves carefully whilst topping up water and dry feed buckets. The dry feed should be kept constantly in front of the calf and should be as fresh as possible. It is essential that buckets are not left empty overnight or for long periods, so they should be topped up at both morning and evening inspection times if necessary.

Digestive upsets are commonly caused by calves eating excessively after a long period without dry feed. They will not eat more than their normal appetite allows, or is good for them at any one time, provided that dry feed is continuously available.

Stale feed, or accumulated dusty feed, should be removed and fed to older stock. Dry feed can be wasted because water has dropped on it from the calf's muzzle, but a vertical strut between the buckets will help to prevent this dribbling. In the observation period the rearer should be looking and listening for signs of bloat which normally occur a few minutes after the liquid feed. This condition needs immediate attention.

Health problems are discussed in Chapter 5, but it is worth pointing out that the post-feeding inspection can be a very useful means of spotting calves which, for example, lie down quickly with stomach ache or are ill in some other way. Once these are dealt with and any necessary fresh bedding given, the calves are best left to settle undisturbed and digest their feed.

If the bedding is clean barley straw, the calves will eat a little quite readily and this can provide all the roughage necessary up to five weeks of age. Calves like routine and should be subject to an unchanging pattern of activity until they are ready for weaning.

Weaning
The approximate time for this can be decided by different means such

as a fixed date from birth (chalked on a pen), or when there is a consistent intake of 1 kg (2.2 lbs) of dry feed per day, or by achieving a suitable weight such as 60 kg (132 lbs) for Friesian heifers or 65 kg (143 lbs) for Friesian bulls.

The weight method is perhaps the least likely to be practised because of the difficulty and time involved in measuring or estimating liveweights. It may be practised subconsciously when dealing with small calves at birth (twins for example). Such calves benefit from an extra period on liquid feed and their progress can be monitored in a few seconds by the use of a weighband pulled tightly round the chest.

It is simple to note when calves are eating 1 kg of concentrates a day by using a chalk mark on the pen for each measure given during the day.

At Barhill where fairly uniform calves are purchased the practice of weaning at thirty-five days from entry has worked well for ten thousand calves reared over eight years. The weaning programme should be according to an agreed policy, but the stockperson should be able to modify this where necessary.

It is important to make sure that bucket feeding is not carried on longer than needed since this adds considerably to feed and labour costs.

Different weaning methods all have their advocates. These vary from abrupt cessation of liquid feeding to a long-drawn-out reduction of both quantity and strength of the milk replacer. A compromise should be the aim; avoiding rapid changes in feeding is always a good maxim. Simplicity is also achieved, for example, on the twice-a-day system by missing the afternoon feed for say, four days. If this can include the weekend, so much the better.

After weaning, calves rapidly increase their intake of dry feed to replace the lost liquid nutrients. The average weekly intake of dry feed in the week after weaning will be 1½ times that of the week prior to weaning. For calves of about 60 kg (132 lbs) liveweight this means that intake increases from 8 kg (18 lbs) in week 5 to 12 kg (27 lbs) in week 6, where the twice-a-day bucket system has been practised.

After weaning, calves can be housed in groups. This should be done ideally one week after liquid feeding has ceased, thereby avoiding two major changes at the same time. Ad lib feeding of concentrates should continue for another five weeks at least for all classes of stock in order to capitalise on the good start made and on the continued favourable feed conversion of 2.85:1 during the 5–10 week period. The early-weaning dry diet fed prior to five weeks is best continued for this period to ensure top performance. The temptation to feed a cheaper lower energy or protein diet should be resisted or performance will suffer. More detail of this is given in Chapter 3.

A Feeding Programme—Mixing Strength
The normal recommended strength is 125 g/litre (2½ oz/pint). The reconstituted milk should be fed at 36 °C (97 °F). Day one is counted as the evening feed for the purchased calf or when the home-bred calf has received four to five days on colostrum/milk. The system is suitable for the use of either milk replacers containing high levels of skim milk powder or for those containing substitutes for milk protein and often referred to as 'zero bag' diets, i.e. containing no skim milk powder. The amounts given in tables 1.2 and 1.3 are for a Friesian calf of 40–50 kg (85–110 lbs) of both sexes.

Table 1.2   Programme for twice-daily feeding *(metric measures, 125 g/litre)*

Days	No. of days	g/calf per day	Litres/calf per day	Litres per feed	Total fed (kg)	
1	1	250	2	2	0.25	one evening feed
2–4	3	375	3	1.5	1.125	
5–31	27	500	4	2.0	13.50	
32–34	3	250	2	2.0	0.75	morning feed only
				Total	15.625 kg	

It must be emphasised that the amounts of milk replacer consumed serve as a reliable guide in line with performances quoted but should be modified to suit circumstances. For example, smaller calves within the Friesian breed or smaller Ayrshire or Jersey calves will have a lower intake and reduced performance accordingly. Or where a group of smaller or late-born calves are concerned, these can be given a longer period on liquid feed to improve performance and bring them nearer to the weight achieved by the rest of the batch reared on normal intakes.

For the main period of liquid feeding, two litres per feed is recommended which supplies 500 g per day of milk replacer powder. In the build-up and weaning periods the same strength of liquid is still used but less is fed.

The amounts shown in table 1.2 are whole litres to aid easy measurement. The mixing strength is constant throughout at 125 g/litre; or as specified by the manufacturer. The total amount fed at 15.625 kg is 25 per cent higher than applies for the once-a-day system.

Table 1.3 Programme for twice-daily feeding *(imperial measures, 2½ oz/pint)*

Days	No. of days	oz/calf per day	pints/calf per day	pints per feed	Total fed (lbs)	
1	1	7½	3	3	½	one evening feed
2–5	4	15	6	3	3¾	
6–31	25	17½	7	3½	28½	
32–35	4	7½	3	3	1¾	morning feed only
				Total	34½ lbs	

The imperial programme (table 1.3) differs slightly from the Metric system in order to keep the pints per feed in whole or half pints. The final quantities are the same, however.

Once-a-day Bucket Feeding

Many rearers find the idea of feeding calves only once a day so illogical that they are never prepared even to give it a try. In any case twice-daily feeding for new-born calves is necessary for a week. Where there are always a number of these in the calf house over the main period of rearing calves from the dairy herd, it is understandable that the switch to once a day is unattractive.

The batch rearer of purchased calves is more likely to look favourably on the once-a-day system since it can be operated after an initial two to four days on twice-a-day feeding which is designed to settle the calves in.

The once-a-day system demands the use of top-quality milk replacer diets for best results since it is beneficial to have the large single meal form a clot in the calf's stomach. This normally requires a milk replacer with a high level of skim milk powder inclusion. The total milk replacer for each day is 400 g, given in one feed of 3 litres (133 g/litre) or 15 oz in 5 pints (3 oz/pint).

Over five weeks this amounts to an intake of exactly 12.5 kg which equates with a 25 kg (55 lb) bag of powder for two calves.

As with the twice-a-day system it is important to feed at a regular time. However the time can be chosen to suit the rearer and this is one attraction of the system. The time most commonly chosen is in the morning to fit in with other work, and to leave the rest of the day clear.

At least one other formal inspection of the calves, other than the single feeding time, must be made to check for illness or accident. If feeding was in the morning this could most conveniently be carried out at tea time. The final quick check at night, which is necessary on all systems, is only to make sure that the animals are safe and secure.

Several trials have been carried out at Barhill where there have been alternate pens of calves in the same house on once-a-day or twice-a-day feeding. Even under these trying conditions, the once-a-day calves soon learn to ignore the afternoon feed of the calves fed twice-daily, demonstrating their attachment to their own routine.

By missing out the feed on one day towards the end of the five-week period weaning is suitably progressed. Some rearers may wish to extend this to two alternate days (e.g. 31st and 33rd) or if the calves are eating dry feed very well, a shorter weaning period may be possible.

The once-a-day system encourages calves to eat dry feed throughout, and Barhill data show a 21 kg intake to five weeks compared to 16 kg where twice-a-day feeding has been used.

Once-daily Feeding Programme—Mixing Strength
For once-daily feeding a higher mixing strength of 150 g/litre (3 oz/pint) is used. Rearers should follow the feeding and cleaning routine already described for the twice-a-day system. The feeding temperature is again 36 °C (97 °F) and day 1 is counted as either being the day a calf is purchased or when a home-bred calf has received colostrum/milk for 4–5 days.

High-quality milk replacers with over 60 per cent skim milk powder are preferred for this system. Tables 1.4 and 1.5 give quantities suitable for a Friesian calf weighing 40–50 kg on day 1 and of either sex.

Table 1.4   Programme for once-daily feeding *(metric measures, 150 g/litre)*

Days	No. of days	g/calf per day	Litres/calf per day	Total fed (kg)	
1	1	200	2	0.2	Plus 1 litre milk for home-bred calves
2–4	3	300	2	0.9	Ditto. Bought-in calves could have two feeds on these days
5–31	27	400	3	10.8	
32 & 34 (miss 33rd day	2	300	2	0.6	
			Total	12.5 kg	

Table 1.5   Programme for once-daily feeding
(imperial measures, 3 oz/pint)

Days	No. of days	oz/calf per day	Pints/calf per day	Total fed (lbs)
1	1	9	3	½
2–5	4	12	4	3
6–26	21	15	5	19½
27–33 and 35	8	9	3	4½
			Total	27½

The total fed is exactly half a bag of milk replacer which makes calculation of total quantities required very easy.

The distribution of milk replacer is not exactly the same in the imperial table as in the metric table (to keep the quantities simple) but the total amount fed is again exactly half a bag. Quantities are all given in whole pints per calf per day for ease of measurement, and a mixing rate of 3 oz made up to a pint applies throughout.

AD-LIB SYSTEMS

Cold Ad-Lib

This system has been practised sporadically in the past and was at one time associated with outdoor rearing. However it was with the introduction of acid milk replacers in 1978 that the system was reintroduced and it rapidly gained popularity, only to be followed by a decline.

The first acid milk replacers were imported from Holland and were of the zero bag type based mainly on whey products. Acid milk replacers based on skim milk powder as well as zero bag products were soon manufactured in Great Britain. In contrast to this country where acid milk replacers are mainly fed from bulk containers, they are mainly bucket fed in Holland.

The addition of acid to milk replacers enabled the reconstituted products to have a life of three or more days without going sour. However, much of the early interest in the acidified cold ad-lib system centred around claims made by users for reduced scouring in calves associated with the acidity of the milk replacer. Although these claims for improved health were not made by the manufacturers of milk replacers in this country, many farmers were convinced that their calves reared on this system were healthier.

There was also some controversy whether mildly acid or highly acid milk replacers were to be preferred. The former were most commonly used with a pH of 5.4 to 5.8. The value of the acidified feed as such cannot be separated from the benefits of the system which leads to calves having a much higher nutrient intake early in life and receiving this on a little-and-often basis by sucking. Observation of the calves when sucking shows a high saliva production accompanied by slobbering. This saliva contains a fat-splitting enzyme which is carried down into the calf's abomasum, thus aiding fat digestion.

It should be remembered in considering the role of acid in the milk replacers which when reconstituted vary from about pH 4.3 (very acid) to pH 5.7 (mildly acid), that the calf's abomasum between feeds has a pH of 2.0 which is extremely acid. Normal milk replacers, so called 'sweet' products have a pH of around 6.3.

Further conclusive work remains to be done on the merits of acid replacers, but three trials at Barhill which compared acidified and normal diets fed to individually penned calves by bucket showed live-weight gain benefits varying from nil in summer to a significant 14 per cent improvement at five weeks in winter when infections and weather were adverse for these bought-in calves.

In group-fed trials run at the same time where calves were given ad-lib access to acidified milk replacer from bulk containers there was a low incidence of illness. The number of rearers practising ad-lib acidified cold feeding has declined somewhat since the initial enthusiasm for the system. There remain a large number of convinced devotees however, many of them rearing home-bred calves as dairy herd replacements.

It may be helpful to list the advantages and disadvantages of the system.

Advantages
1. It is a simple system and can be labour-saving if well organised. In particular the workload can be conveniently arranged with a minimum amount of work over weekends, and a possible saving in overtime.
2. Contrary to some prophecies of doom and a few well-publicised failures, the system does actually work well and there appear to be possible health benefits associated with it.
3. Calves grow and thrive well. Moreover they show a significantly improved bloom over bucket-reared calves at five weeks, which pleases the rearer and is indicative of a calf growing well and using nutrient intake efficiently. Where bull calves are sold the extra bloom enhances the price received.

Plate 2   Cold ad-lib system: simple arrangement with minimum equipment

4. There is a low requirement for equipment, allowing a capital saving which is particularly attractive for the rearer starting without specialist buildings or needing to expand.

Disadvantages
1. It is principally criticised for its higher feed cost compared to restricted intake on conventional bucket-rearing systems. The extra feed cost is about £25 to twelve weeks for which there is an extra gain of about 7 kg which is worth a minimum of 70–80p/kg, i.e. £5.50.
2. Extremes of weather can cause practical difficulties. In prolonged sub-zero temperatures it may not be possible in unprotected housing to prevent the pipelines freezing up. Conversely in very hot and 'muggy' weather some milk replacers will turn sour much more quickly than normal.
3. Because of the high intakes per calf a lot of liquid feed has to be mixed up and taken to the point of consumption. Inevitably there is a high urine production. Well-drained and adequately bedded pens are necessary.
4. Some would say that a higher degree of stockmanship is required to notice signs of illness in individual calves. Moreover infections can be more easily spread where calves are group housed.

The decision to choose the cold ad-lib system is therefore very much an individual rearer's choice in his own circumstances and for his own objectives.

Equipment Needed
Although some advisers advocate single penning of calves for a few days to ensure that they have taken to the system, this is counter-productive to the goals of minimising equipment and labour. Group housing is therefore recommended for both home-bred and purchased calves. From the start up to twenty calves per pen can be reared provided there are sufficient teats. Not less than one teat per seven to eight calves should be available. Of course, if desired, more than one pen can be supplied from a single barrel strategically placed.

Large-diameter soft teats with a self-closing hole are preferred. Calves are able to form a good seal with their tongue when sucking on this type of teat and the closing of the hole prevents the back-flow of liquid when sucking ceases. If they are mounted in a screw-on holder it is very easy to remove them during the weaning phase and reduce damage from chewing.

The teats should be mounted on a board pointing downwards at the natural angle of the calf's upturned head. The height above floor level depends on the size of calf but is usually about 600 mm (2 ft) with 1 m

(3 ft) between teats. Plastic tubing leads from the bottom of the bulk container. It should be long enough to curl and lie on the bottom with the aid of a weight such as a piece of metal tube.

A one-way valve can be inserted in the end of the tube to prevent back-flow of the liquid. This is most helpful in the learning phase before the calves are sucking strongly. The valves are marked to indicate the direction of flow and need regular cleaning to keep them effective in use.

A minimum of 1.5 sq m (16 sq ft) per calf is necessary within the pen. These can be formed of home-made wooden hurdles, removable for easy mechanical cleaning-out. For easy access to fill the containers it is important to form a passageway within the building. A 200-litre (45-gallon) container is a suitable size and holds a day's supply of milk replacer for sixteen to twenty calves at the maximum likely intake of 10–12 litres. It is very important that the container has a tight-fitting lid to keep out flies and dirt.

The pens should also have a self-filling water bowl or water bucket and a rack for hay or straw. Concentrates should be on offer from the beginning of the rearing period in a trough placed in position to avoid fouling and so as not to form an obstruction on which the high-spirited calves can easily injure themselves.

In order to meet the requirement for minimum labour, a mechanical mixer on wheels is a great help. Rapid filling can be arranged from a header tank fitted with a large-bore outlet. If the tank is fitted with a heater element and low-level thermostat, the water in winter can be instantly available at not less than the minimum of 10 °C (50 °F) recommended. If water colder than this is used, the intake per calf and subsequent performance are depressed.

If there is scope in planning a new or converted shed for ad-lib rearing, adequate space for feed storage and a work area should be allocated. A washing-up trough and storage cupboard complete the simple equipment requirements.

Feeding

The rate of mixing of milk replacers for ad-lib feeding recommended by most manufacturers is 125 g/litre (2½ oz/pint). The mixing is usually carried out using 'cold water' except for the first two days when purchased calves are being taught to suck. Small quantities of a warm mixture are very helpful at this training stage. Whenever calves are introduced to the teat for the first time, whether coming direct from the cow, or purchased, they should be held to the teat and encouraged to suck by pressure from a hand round the muzzle. Failure to make sure that each calf was able to obtain adequate feed intake was a frequent cause of poor results in the early days of the system.

Most calves learn to suck immediately and can be marked accordingly with a coloured crayon. The few difficult ones need perseverance but they will eventually learn. The fullness of these calves in particular should be observed, because they may be sucking the teat perfectly satisfactorily when the stockman is absent.

Once a group of calves is established on feed the aim is to keep a supply of milk replacer in front of them all the time, but allow the containers to become empty every few days for cleaning. The exact timing of this has to be left to the operator, and depends very much on the ambient temperature used. Generally it will vary from two days in hot weather using a mild-acid milk replacer, to five days in cold weather using a high-acid powder.

As a guide to quantities, calves will take about 6 litres (1½ gallons), 8 litres (1¾ gallons) and 10 litres (2¼ gallons) per day for weeks 1, 2 and 3 respectively. For example a pen of ten calves on the top rate of intake will need a mix of 100 litres (22 gallons) every other day. Cleaning would take place therefore every second and fourth day depending on the weather, in general avoiding cleaning at weekends.

At this rate of consumption (which may be less on the very acid powders because of their lower palatability) the total powder fed by the end of three weeks will be about 20 kg (44 lbs) which is high compared with 8 kg (17½ lbs) on once-a-day bucket feeding. Conversely the quantity of dry feed eaten will be negligible because the calf is able to satisfy its appetite in liquid form.

Weaning
It will be clear that calves cannot be allowed to go on consuming milk replacer at the rates quoted which reach 1.25 kg (2¾ lbs) per calf in the third week. The system therefore has to change from ad-lib cold to restricted cold in order to limit intake of milk replacer and to stimulate dry feed intake as an alternative. Where a whole pen of calves is of approximately the same age and size and is to be weaned as one unit, this makes management easy. They simply stay in the same pen until the weaning phase is complete.

On a small farm it is often convenient to have one ad-lib pen for dairy heifer replacements, adding new calves as born and removing others at about three weeks when they are due to start the weaning process. These calves go into a weaning pen and receive restricted amounts of milk replacer for two weeks.

Periods of up to one week have been found to be inadequate to prevent a weaning check and at least two whole weeks up to a weaning age of five weeks is strongly recommended. During this period of restriction a further total of 10 kg (22 lbs) of milk replacer powder is allocated on a declining scale.

Opinions vary on the best way to do this. The simplest way, with the best control, is to put in 8 litres (1 kg powder), 5 litres (0.63 kg), and 3 litres (0.38 kg) per day for each calf for successive five-day periods. The total intake of powder is then 10 kg (5 + 3.1 + 1.9). Imperial quantities would be 1¾ gallon, 1 gallon and ⅔ gallon for each five-day period using a consistent strength of 2.5 oz/pint.

Perhaps an even simpler method and routine is to unscrew the teats for a period each day. The time allowed for feeding could be a fixed one of say five hours throughout the whole two weeks or a declining one arranged within normal working hours. Some rearers change the supply tubes from milk replacer to water, but this is not a good practice. It encourages the calf to remain attached to the teat instead of learning to drink its water and eat dry feed.

The restricted sucking period will be a rough and tumble affair with a great deal of competition for the teats when they are put on. However in the period allowed there is ample time for all calves to have a suck, and those not so inclined will be forced more quickly on to the dry feed. A highly palatable 18 per cent crude protein dry feed should be on offer right from the start of rearing on this system even though amounts taken at first are very small.

It is particularly important that the dry feed has adequate protein, especially undegradable protein (UDP) for at least four weeks after weaning. Once weaning is complete the calf loses its supply of milk protein which on a sucking system was taken directly into the abomasum. It is then wholly dependent on dry feed for its protein needs.

Amounts of dry feed eaten on this system will be approximately 1.5 kg (3½ lbs) to three weeks and 8 kg (18 lbs) to five weeks. After weaning is completed dry-feed consumption will increase rapidly reaching 3 kg/day (6½ lbs) by ten to twelve weeks, provided that no restriction is applied.

If the calves can be left in their weaning pen for a week or two until they are eating well they will settle better than if they are removed to strange surroundings.

Machine Feeding System—Warm Ad-Lib

Many similarities between the cold ad-lib and the machine warm system exist, because the calves are group housed and are on a high plane of nutrition on both. Although calf-feeding machines have been available for many years, the number of calves reared in this way until 1979 was only 10–15 per cent of the total.

Currently there is a greater interest in machine rearing and there is a willingness to spend £800–£1400 on a machine to rear calves. This is especially true in large dairy units where there is limited labour available for calf rearing. It is probable that saving as much labour as possible is the main reason for the renewed interest.

The system has the advantages of group housing and frequent feeding as with the cold ad-lib method, but without the chore of having to mix and carry liquid milk replacer around—the machine does all that.

Most manufacturers now sell special milk replacer powders for machine feeding, and some are linked with the sellers of the machine. The main requirement is that the powder flows consistently from the machine's hopper into the mixing chamber. If the storage vessel for the reconstituted liquid—either heated or just insulated—has no agitator it is essential that the milk replacer used does not produce a sediment. Most milk replacers in this situation would contain over 60 per cent skim milk powder. If however there is a stirrer or agitator in the storage vessel it may be possible to use milk replacers containing other animal or vegetable proteins which tend to settle out. Since these give economy in formulation they can partially offset the higher consumption and feed cost associated with this system. Machines which have several pipeline feed points can, of course, service two adjacent pens, so that one of these can be on a 'weaning regime' (as described for the cold ad-lib system), whilst the other remains on full 24-hour access. Rather than just disconnecting the pipe for particular periods of time during the two-week weaning phase, it is again better to unscrew the teats which only takes a moment. This saves wear and tear on the teats and encourages the calves to transfer their activities elsewhere.

One idea for promoting the intake of dry feed is to put the feed trough well away from the machine and to illuminate it at night, leaving the machine in near darkness.

The temptation to wean home-bred calves abruptly is probably greater when machine rearing compared to the cold ad-lib system. It is very simple just to switch off the machine, and this may have much to commend it, as long as the calves are not under your bedroom window! Since the inevitable check which results is unlikely to be monitored on a weigh scale it would often be unnoticed. In any case the calves have had a good start and are well grown for their age; the check is perhaps not important in the overall growth pattern of a dairy heifer being well treated to calve at two years.

It is important to check on the strength of the mixture being dis-

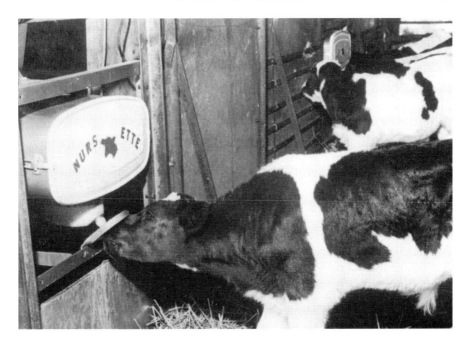

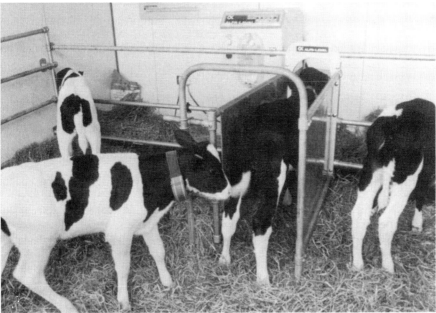

Plates 3 and 4   Two machine feeding systems. The lower photograph shows neckband transponders being used to allocate quantities of feed to individual animals

pensed by the machine. Unfortunately the controls for this are some-times not very precise with no convenient scale. Adjustments have therefore to be by trial and error. The mixing strength should be in the normal range of 100–125 g/litre (2–2½ oz/pint). To calibrate the machine use a paper formed into a cone or a carton to catch the dry powder as it is dispensed. It can then be weighed. The water dispensed at the same time can be poured from the mixing vessel and measured and the necessary calculation of the strength of mix made.

Some of the more sophisticated machines have delay mechanisms so that calves cannot suck continuously for prolonged periods. If these are associated with recognisable noises or lights all sorts of 'Pavlovian' responses can be set up. When there is no restriction of supply, other than the ability of the machine to keep up, there is usually competition at the teats with calves pushing each other aside to gain possession. Consequently any one calf is prevented from having a 'solo run'.

Further developments are taking place in computerised rationing on the machine by means of neck collars and transponders as used in out-of-parlour cow feeders. This would provide ultimate control of desired intakes where this is thought to be necessary.

As with the cold ad-lib system, a generous pen area with plenty of bedding helps to counteract the high urine output and keep the calves clean. On this system it is obvious how lively the calves are. Much running about and group activity alternates with calves 'sleeping it off' in a prostrate position which can cause anxiety to the stockman at first.

The pen should be free of any projections which could cause serious leg injuries to the calves, especially if they move quickly when startled. A quiet entry to the calf house should always be made, and it will be noticed how friendly and inquisitive calves are on this system.

Apart from routine care of the machine including a daily clean of the storage vessel and pipes, there is little to describe in the management area. It is easy to check on each visit to the house that the powder is topped up and flowing and that the machine is dispensing properly. Preparations for a prolonged failure of either electricity or water supply need providing for, probably by putting the supply pipes into a container of mixed warm-milk replacer.

CHOOSING THE APPROPRIATE SYSTEM

Having described the main systems of calf rearing, each can be said to have its own advantages and disadvantages. There certainly is no single 'best' system and each individual rearer will make his own choice to suit his circumstances. They all work well and can be used to rear first-class calves.

Detailed costings are given in the Appendix, but the most recent costings (1995) show that the systems do differ in feed cost per kg of liveweight gain, with the ad-lib systems costing more.

Table 1.6 Feed costs per kilogramme of liveweight gain under different systems

0–12 weeks	Bucket once-daily	Bucket twice-daily	Cold ad-lib	Machine warm ad-lib
Cost per kg liveweight gain (*pence*)	86	92	114	114
Cost as a percentage of the least cost	100	107	132	132
Feed cost per calf (£)	51.59	55.40	78.35	78.35

These costs are based on accurate records of feed intake and liveweight gain for several thousand calves. They will of course vary in practice from one farm to the next and as feed prices change. However the relative costs between the systems will remain similar.

The main factors which determine the feed cost per unit of gain are the intakes and cost of milk replacer and dry feed together with the liveweight gains made. An outbreak of illness can have a major effect on efficiency of feed utilisation and cause a group of calves to have a very high cost per kg gain. Similarly, if the calves are castrated in the 5–12 week period, liveweight gain and feed conversion will be checked.

Moreover the average figures shown, although reliable because they are drawn from large numbers of calves reared, do conceal quite large variations between individual calves and between batches.

The costs shown above are based on milk replacers at £1500 per tonne (£12 higher for the acidified product) and £210 per tonne for the dry feed. These prices are typical of the premier products on the UK market but do not take into account discounts for large users and for taking dry feed in bulk, both of which will reduce the feed cost per calf.

The costs are also based on strict five-week weaning of purchased Friesian bull calves which is a well-proven system. However surveys have shown that about 40 per cent of rearers continue liquid feeding to six, seven or even more weeks. Not only does this increase the labour input, but it also increases the feed cost of liveweight gain. This is

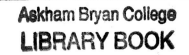

because the cost of energy in milk replacer is 4 times the cost of energy in dry feed. Having said that, there is no precise 'correct' time to wean, or a 'correct' amount of milk replacer which should be fed. The optimum will vary from farm to farm, depending on management objectives and facilities. The specialist rearer will tend to wean earlier than 5 weeks when the calves are doing well in order to reduce feed costs. The costs above were based on milk replacer and dry feed intakes as shown in table 1.7.

Table 1.7   Milk replacer and dry feed intakes

		Once-daily	Twice-daily	Ad-lib cold	Ad-lib warm
Milk replacer intake to five weeks (kg)		12.5	15.5	31	33
	(lbs)	28	34	68	73
Dry feed intake to twelve weeks	(kg)	154	150	132	135
	(lbs)	340	330	290	298

The amounts of milk replacer on the bucket system are fixed by the recommended daily timetable, but the intakes on both ad-lib systems are determined in the first three weeks by the appetite of the calves.

Typical daily intakes on the cold ad-lib system would be an average of 0.75, 1.00 and 1.25 kg (1.65, 2.2 and 2.75 lbs) for weeks 1, 2 and 3 respectively. On the machine, intakes tend to be slightly higher in the early stages. The diagram in the Appendix shows a comparison of the intakes on the different systems week by week.

One example where the flying start given by the ad-lib system would be appropriate is for intensive bull beef. These need to complete their growth within a year gaining about 380 kg (840 lbs) which requires a performance of 1.2 kg (2.6 lbs/day) liveweight gain overall. Economic gains in early life would be made by such bulls converting feed at 3:1 compared with a conversion of 6.5:1 just prior to slaughter.

Similarly for the early-autumn-calving heifer with an overall growth requirement of 0.65 kg (1.43 lbs) per day to achieve a calving weight of 500 kg at two years, good growth rates early in life are certainly efficient. They also give an optimum liveweight at turnout and allow her to make the most of the first summer's grazing, and to utilise a high proportion of roughage in the second winter. In saying this it is assumed that turnout management is good with concentrates being reduced before turnout and continued at 1–1.5 kg per day at grass for up to two weeks with hay or straw also on offer. The object of this is to adapt the calves' digestive system to the lush spring grass which is high in both energy and protein. As such it needs balancing with fibre from the long fodder and possibly also by having a high fibre concentrate.

By contrast to these two examples the specialist rearer producing a reared calf of 100 kg liveweight for sale at eleven to twelve weeks will be looking for the lowest feed cost per unit of gain and will therefore favour once-a-day bucket feeding. Currently (1995) the rearer might be paid £80 to rear a calf to 110 kg LW and 70p per kg above that. He needs therefore to keep all his costs down to show a margin. As indicated in the costings, labour and depreciation often form part of this margin.

Chapter 2

GENERAL MANAGEMENT OF CALVES

HOME-BRED DAIRY REPLACEMENTS

THE REARING of heifers for herd replacements is a relatively costly item in the dairy enterprise accounts. It is non-productive in the sense that it does not provide a cash income and, for that reason, a dairy farmer with limited acreage may opt for buying in his replacements at first calving.

In round figures the well-bred purchased heifer costing £1000 could produce milk worth a similar cash value in ten months (say 4,200 litres at 24p per litre). If, in subsequent calvings, she also produces beef-cross calves which are usually worth £40–50 more than the pure Friesian calf, the attraction of this policy becomes apparent.

Not everyone can follow this path, of course, or there would be no replacements. And moreover, buying in calving heifers can be full of pitfalls for the unwary. The rearing of heifers on good land (compared to the use of hill areas where the land is not suitable for dairy cows) demands an intensified system. This involves high stocking rates and early calving in order to make the land use competitive with alternative enterprises.

Now that tuberculosis and brucellosis are controlled any health advantage of rearing one's own replacements is less dramatic, but there is still the benefit of a controlled health programme which covers husk and intestinal worms, summer mastitis and warble flies.

Perhaps less obvious, but even more important, is the matter of stock training to produce docile animals. This development of a bond between stockman and calf starts at a very early age and remains after the heifer joins the milking herd. The Americans describe this bond as that of the animal having a small 'flight distance' and reckon it to be very important for the highest-producing animals. Then, for many farmers, the strongest reason for rearing their own replacements is the desire to have a programmed breeding policy and to make it possible to create a distinct herd rather than a collection of cows. This is a long

30

Plate 5   Confident bonding between heifers and their rearer

process, often a lifetime's work, but the financial rewards and sense of personal satisfaction can be great.

Genetics as a science can now be applied with greater effect in the dairy herd than ever before, thanks to the computer and sophisticated bull progeny testing schemes. However, the eye of the stockman remains extremely important in selecting desirable conformation characteristics and temperament as part of livestock improvement. These are of as great consequence for the high-yielding long-lived dairy cow as milk yield and quality, and make up a large part of the analysis of the progeny of a dairy bull.

If a Friesian dairy heifer enterprise needs to be intensified, the most effective policy that can be pursued is to calve down autumn-born calves at two years, and spring-born calves not later than two years six months. (Reducing the age at calving will also speed up the programme of genetic improvement.) And if a higher stocking rate can also be achieved, this will further improve productivity, provided the area released is profitably utilised for more cows or alternative cropping. The actual number of heifers needed is determined both by the replacement rate as well as the age at calving, as table 2.1 shows.

Table 2.1   Number of heifers of all ages being reared per hundred cows

		Age at first calving	
Replacement rate	2	2½	3 years
20 %	40	50	60
25 %	50	62	75
30 %	60	75	90

There is plenty of evidence from ADAS, MMB, NIRD and practising farmers that calving Friesians for the first time at two years is easily achievable and that growth targets are very moderate with no risk of 'spoiling' the heifer.

Table 2.2   Suggested target liveweights for two-year-old calving

	Ayrshires				Friesians			
	Liveweight		Gain/day		Liveweight		Gain/day	
	kg	lbs	kg	lbs	kg	lbs	kg	lbs
Birth	34	75	—	—	43	95	—	—
3 months	75	165	0.45	1.0	93	205	0.55	1.2
15 months (bulling)	270	595	0.53	1.2	320	706	0.63	1.4
24 months (pre-calving)	430+	948+	0.59	1.3	510+	1,124+	0.70	1.5
Overall growth rate	—	—	0.54	1.2	—	—	0.64	1.4

Recent work at the Institute for Research in Animal Diseases, Compton, has shown that a gain of 0.8 kg/day up to the age of puberty, about 230 kg (506 lbs) for a Friesian, is perfectly satisfactory, but that a gain of 1.0 kg per day up to this stage results in lower milk yields in the first and subsequent lactations.

An ADAS survey on difficulties at calving (dystocia) showed only a 10 per cent incidence for heifers calving at 2–2½ years, which was the lowest of the various ages, as shown in table 2.3.

Additionally an NIRD survey (table 2.4) has shown that there are fewer problems getting heifers in calf at an earlier age.

Table 2.3   Difficult calvings related to age at first calving *(ADAS)*

Age in years	Number calved	% calving difficulty	% calf mortality
Under 2	277	16	12
2–2½	768	10	11
2½–3	345	13	12
Over 3	60	22	28

Table 2.4   Conception rates for heifers *(NIRD)*

| | Age at first calving | |
	2½–3 years	2 years
Number of heifers	202	145
Conception rate to first service	55.4	69
Services per conception	1.54	1.34
Percentage conceiving to all services	88.6	94.5

Some dairymen will perhaps neither have land or feed good enough for calving at two years, nor believe it is the optimum age. They will point out the MMB figures for heifer lactations linked to the age of calving which show a reduction of 40 kg (8½ gallons) of milk for each month below three-year-old calving. However this is for all heifers as they appear in the statistics and is not what will happen when two-year-old calvers are properly managed. The yields shown in the statistics, moreover, are more a reflection of size than age, and although the first lactation yield may be lower, the herd life and lifetime production of two-year-old calving heifers is higher than for later calvers.

Table 2.5   Effect of age of first calving on lifetime yield *(MMB)*

Age at first calf	Herd life no. of lactations	Lifetime production (litres)
2	4.0	18,708
2½	3.84	17,927
3	3.78	17,621

A guide to the actual level of production in the heifer lactations to be expected comes from the Barhill herd where, over the last ten years they have been 82.4 per cent of second and subsequent lactation yields. The averages have been 5,099 litres and 6,181 litres for heifers and cows respectively.

We are concerned here with the early rearing stage, but it is worth emphasising that the heifer at two years weighing 500 kg or just over before calving is still three or four years away from her mature weight and therefore she must be well fed for milk production and for continued growth in her first lactation especially.

MMB data show that the majority of Friesian heifers are calved during the autumn. Most of these will have had two whole summers at grass, and when first turned out at six to seven months of age the minimum target weight should be 180 kg (400 lbs).

If this weight is reached in six months overall, the gain will have been 0.75 kg (1.65 lbs) per day, but if reached in seven months only 0.65 kg (1.4 lbs) per day. Late-born calves should be made to wait before being turned out and will probably need a little trough feed all summer in a group on their own.

The main group of calves weighing 180 kg or more at the time of turnout will, after an initial period of acclimatisation, make adequate gains of 0.7 kg per day on good grazing alone. Indeed for the next eighteen months they need only a small amount of concentrate supplementation, provided the forage available is of good quality. This should be checked just as carefully as for dairy cows, and the progress of the heifers monitored, particularly at the critical time coming up to service. Where a weigher is not available this can be done by using a weighband which is quite accurate enough for this purpose and is simple to use.

Many data from Barhill, with both Friesian and Hereford x Friesian beef heifers, have shown that the better-grown the autumn-born calves are when they are turned out, the better use they make of summer grazing. Thus, within any grazing group, the best-grown calves draw ahead and the smallest calves tend to become victims of any grazing or managerial deficiencies.

However for high gains at grass it is essential that the winter feeding is based on a high proportion of good-quality roughage, and that the change-over to grass is well cushioned by a gradual withdrawal of winter feed after turnout. The level of concentrate feeding through the winter can be at a constant average of 2 kg (4½ lbs), varying 0.5 kg up or down according to the quality of the roughage. The increasing appetite of the growing animals will be met by the roughage, which should be available on free access.

Plate 6   Ad-lib silage feeding between 12 weeks and turnout

The greatest hazards facing the grazing heifers in their first year are from lungworms (husk), and intestinal worms. Effective means of control of these will be detailed in Chapter 5. An important part of this control is management of the grassland to provide adequate nutritious young grass, alternating the heifers with sheep, or cutting for hay and silage.

To return to the calf-rearing stage of rearing heifer replacements, with the assumption that early calving is worthwhile, one is often asked which system is the best.

The answer must be that bucket rearing, ad-lib cold, or machine feeding are all very suitable and effective methods of rearing. Cost per kg of gain to twelve weeks will vary, however, from a high of 114p on machine feeding down to 86p on once-a-day bucket feeding. This difference is an important consideration amounting to £21 a calf in feed cost reared to the same weight.

Time available to the stockman, normally hard pressed at the busy calving peak, is certainly becoming less rather than more and this favours machine feeding in particular. If this can be combined with simple penning of calves in groups of six to ten with easy mechanical cleaning-out of pens followed by thorough disinfection, then advantages will come from healthier calves and lower mortality. An additional

Plate 7 (*Above*) Guiding the new-born calf to the teat to ensure adequate colostrum

Plate 8 Giving colostrum or saline solution by means of stomach tube

Plate 9 (*Above*)
A communal colostrum
system

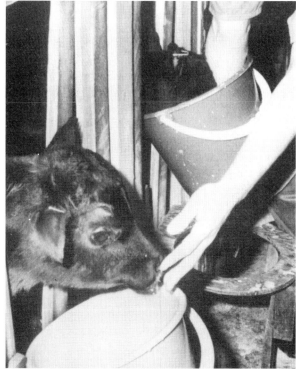

Plate 10   Teaching the
calf to drink from a bucket

pressure is created for the herdsman when Friesian bull calves and beef breed crosses are making good prices. With £140 for Friesian bulls and £200 plus for beef crosses being paid (1995) there is much to be gained from putting plenty of flesh on the backs of calves before sale. This is most easily and effectively achieved by ten to fourteen days of machine feeding.

A combination of systems is, of course, possible. On many farms the herd size has increased substantially over the last ten years. Actual average numbers of cows per herd was thirty-eight in 1973 and is now over seventy. Often accommodation for calf rearing has not increased along with new cow accommodation so that some improvisation has been necessary. In the early part of the autumn, when the weather is not severe, a machine can easily be installed in a covered yard to rear calves in what is normally their follow-on accommodation. The calf shed proper can come into use later, thus avoiding a build-up of E. coli organisms which so commonly reach a peak after the New Year causing scouring and ill health.

It is known however from survey work that most farmers use only one feeding system and that bucket rearing is still by far the commonest. Those who bucket rear are also mostly using twice-a-day liquid feeding for home-bred calves. This is understandable because the animals in the calf house will often be of mixed ages with the younger ones needing two feeds per day. Perhaps it is worth emphasising that calves soon adapt to once-a-day feeding, even when younger calves in the same shed are being fed twice daily.

Management of the Home-bred Calf

Every calf rearer must be aware of the importance of colostrum to the newborn calf in helping to protect it from disease. Immunity which is passed from the cow to its calf via the colostrum is specific to the range of organisms found on that farm and only lasts for a few weeks until the calf can build its own defences.

In spite of this awareness, measurements of the immune globulin levels from colostrum in the blood of large numbers of purchased calves shows about half of them with poor or very poor readings. This indication of low colostrum intake probably extends to heifer calves staying at home. To check this your veterinary surgeon may recommend running a test on a series of calves and this may prove to be salutary for both herdsman and farmer. One simple test, using the radial diffusion kit, only requires a drop of blood from the calf's ear and is easily done. After incubating the prepared plates to which the blood has been added, the immune globulin levels can be read off and expressed as mg/ml of Ig G. A satisfactory level is 20 mg or more and this correlates well with the zinc sulphate turbidity test in ZST units which is the original standard method of measuring immune globulin status.

Of great practical help in explaining why calves do not all reach high levels of colostrum intake was a continuous observation survey carried out at Reading University. Various causes were pinpointed for the failure of the calf to suck a significant amount of colostrum from its dam in the vital first few hours of life. Some calves were lethargic and needed to be moved into action by the stockman, especially during cold weather. Some heifer dams were too excitable or 'fidgety' to allow their calves to suck. The most common cause seemed to be the difficulty some calves had to find and hang on to the teats of large udders close to the ground. This was a feature of older cows.

What can the stockman do to help? A few minutes guiding the calf to the teat may be all that is required, aided perhaps by a small feed for the dam. If no progress is made this way, some colostrum milked from the dam can be given to the calf via a drenching bottle or stomach tube.

A communal system which continues to give calves the maximum of available colostrum for three or four days, with minimum of effort, is easy to set up. A plastic container to hold about 50 litres (10 gallons) of colostrum/milk is fitted with up to four soft teats with self-closing orifices. This should be sturdily mounted on a stand in the centre of a small pen giving easy access to six to eight calves. The container is replenished with colostrum/milk from newly calved cows twice daily and the calves allowed ad-lib access. Each calf needs an initial introduction to the teat and the vast majority of calves need handling only once. Occasional calves, as with any hand-feeding system, may need extra care and attention.

Relatively large amounts of liquid are consumed—up to 8 litres (2 gallons) a day and the pens need generous bedding and good drainage. An ideal arrangement is to have two pens for alternate use and cleaning. Calves can go from this pen on to any rearing system thereafter, either ad-lib or on the bucket, having had a flying start.

Dry Feed Intake

In Chapter 1 the important point was made of letting calves have access to dry concentrate feed throughout the pre-weaning period. This is worth re-emphasising, and the wide variation in intake between calves fed on the bucket system is worth noting also. Typical intakes of 15–20 kg (33–44 lbs) to five weeks are found depending on birth weight and the level of feeding of milk replacer. As a rough guide, intake of dry feed to five weeks will increase by 1 kg for each extra 1 kg in birthweight, but the relationship is not an exact one.

This is probably why the rearer with relatively few calves, giving individual attention, can achieve above-average results. By keeping the dry feed fresh and palatable, and by encouraging the calf with a little hand feed after drinking its milk replacer, the rearer can teach his calves to eat well and to begin ruminating early.

As you might expect there is a very good correlation between the amount of dry feed eaten to five weeks and the liveweight gain over the same period. Each extra 1 kg of feed consumed results in 0.5 to 0.6 kg of extra liveweight gain. Such a good performance ratio will never be achieved again in the animal's lifetime. Anything that can be done in formulating a dry feed to make it more palatable must be beneficial, and this will be considered in Chapter 3.

Post Weaning

Once milk replacer feeding stops altogether the calf must obtain all its nutrient intake from dry feed and this should also entirely satisfy its appetite. Bucket-reared calves will make the change at weaning with a minimum of fuss. Records from a large number of calves show an increase in intake of 50 per cent in the week after weaning compared with the week before where a palatable concentrate feed is offered ad-lib.

Ease of management, freedom from digestive upset, and an economical feed conversion of 2.8:1 all favour ad-lib feeding in the post-weaning period from five weeks. The same high-quality early-weaning concentrate containing at least 16 per cent protein should be fed during this period to compensate, as far as possible, for the loss of milk nutrients. At ten weeks when the calf should have doubled its birthweight, a decision on level of feeding can be made in the light of the quality of roughage available and the desired performance.

Heifers or steers intended for summer grazing will normally receive 1.5–2.5 kg (3.3–5.5 lbs) of concentrate feed for the rest of the winter. Its protein component should not fall below 14 per cent and may need to be 16 per cent with poor hay or silage.

Steers or bulls for intensive feeding will remain on ad-lib concentrates which will gradually be reduced in protein percentage as intake rises. Opinions differ on exact levels, but for steers 12 per cent protein is thought to be adequate, whereas quicker growing and leaner bulls probably need 14 per cent protein.

Although beyond the scope of this book, it is worth saying that the calculations of target growth rates, using available feeds, are easily made for both the first and second winters.

Tables of dry matter intake are very reliable as a guide to total appetite and the amount of energy and protein supplied within that appetite will give a known rate of gain. At present the best source of working data is the Ministry of Agriculture's *Reference Book 433* in relation to the Metabolisable Energy system used for calculating requirements. This is in the process of being updated according to the revised Agricultural Research Council's standards published in 1980, and it will then contain requirements for protein. These will be expressed in terms of the new protein system for ruminants of rumen degradable protein (RDP) and undegradable protein (UDP). The examples shown in tables 2.6 and 2.7 will help to make this clear.

Table 2.6   Requirements for a heifer weighing 175 kg (385 lbs) to grow at 0.75 kg (1.65 lbs) per day
Ration 1—Silage, straw and compound

Fresh weight kg/day	Feedstuff	Analysis			Supplied		
		DM %	ME MJ/kg DM	DCP % DM	DM kg	MJ	DCP g
7.2	Moderate silage	25	9.0	10	1.8	16.2	180
1.4	Barley straw	86	6.0	0.8	1.2	7.2	10
1.9	Compound 14 % CP	86	12.5	11.5	1.6	20.0	184
				Total supplied	4.6	43.4	374

Dry Matter (DM) appetite = 4.6 kg (10 lbs) per day. Energy for Maintenance and growth = 44 Megajoules (MJ). Protein requirement = 375 g of Digestible Crude Protein (DCP) on the old system or 310 g RDP plus 50 g UDP on the new system.

The suggested amounts are sufficient to meet both the energy and protein requirements of the heifer with the protein shown in DCP terms. (RDP and UDP requirements are also met but calculations are not shown.)

In practice it is likely that the farmer would interpret this calculated ration by feeding 2 kg of a 16 per cent compound nut or equivalent concentrate mix, together with silage judged by the forkful and with ad-lib barley straw in the rack. Incidentally spring barley straw, though not so plentiful these days, is preferred to winter barley straw because it has a higher digestibility and is more palatable.

An alternative ration is shown with good-quality hay, plus a home-mix of three parts oats, two parts sugar beet pulp and one part of a 34 per cent protein balancer.

Table 2.7 Ration 2—Hay and home-mix

Fresh weight kg/day	Feedstuff	Analysis			Supplied		
		DM %	ME MJ/kg DM	DCP % DM	DM kg	MJ	DCP g
3.0	Good hay	84	8.5	6	2.5	21	150
2.3	Home mix 14.5% CP	86	11.5	11.5	2.0	23	230
				Total supplied	4.5	44	380

Roughage

I suppose most of us agree with the old adage 'Good hay hath no fellow'. Because it has a wonderful smell and because dairy cows will 'break their necks' to get hold of it, we are conditioned to think of it as the ideal food for ruminants. And so it is in many respects.

Unfortunately the advice given in the past to save the very best 'calf hay', often fine, soft and full of herbs, for the calves, may not be in their best interests. In the first few weeks of rearing and certainly during the pre-weaning stage, the rumen of the calf is not developed enough to digest large amounts of fibre by bacterial fermentation. Consequently, when offered very good hay which is attractive and palatable, the calf consumes more of it than it can easily digest and as a result becomes pot-bellied from distension in the rumen.

The hay itself, however good, contains less energy than early-weaning concentrates even if it is properly digested so that if hay consumption is increased and concentrates reduced, growth will be slower than required. A useful alternative, now adopted by many rearers, is to offer good clean barley straw, either in a rack or, up to weaning, simply as fresh daily bedding. The calves looking for a little long fibre will take the straw when only a week or two old. The total eaten by five weeks will, however, be less than half the quantity of hay normally eaten by that age.

From about ten weeks of age the calf will be able to digest fibrous feeds more efficiently in its rumen and absorb the end products of fermentation through the rumen wall. A change then to hay or silage of good quality will help to meet increasing appetite as economically as possible from this stage onwards.

THE MANAGEMENT OF BOUGHT-IN CALVES— THE SPECIALIST CALF REARING UNIT

The calf unit buying in calves will most commonly be rearing them for beef production, either taken right through on that farm or, more often, sold as reared calves weighing about 110–120 kg (240–264 lbs). The principles of calf rearing discussed for home-bred calves apply just as well for this system, although some practical differences are important.

Disease control, with calves being mixed together from many different sources, is more difficult initially. However there is an advantage with batch rearing of purchased calves in that thorough cleaning of the house and pens is possible between each intake and this gives good hygiene. The phrase 'all in, all out' borrowed from the poultry industry is a fundamental requirement for the specialist rearer. Much will then depend on the buying skill of the rearer or his calf buyer. If future legislation prevents the widespread movement of calves from market to market, this would be of immense benefit to the rearers in reducing the incidence of disease.

The economics of the rearing enterprise are more critical and there is very little opportunity to counteract management mistakes by the time a calf is ready to be sold again. The specialist rearer needs to know his exact costs and returns on each batch and his management needs to be of a high standard.

In rearing a batch of calves for beef production there is an advantage in having a uniform group of calves which can be managed as a unit. Initial health problems and the build-up of immunity to disease tend to occur to the batch as a whole.

Buying the calves on any scale will normally be best carried out by the skilled calf dealer who can be instructed on sex, breed, type of calf and weight range required. On the other hand the smaller-scale rearer can often pick up a 'bargain' by being in the right place at the right time. He has to be prepared to put his group together over a week or two however, and has to have time available to attend the auctions.

Where the reared calves are to be sold at 110 kg or more, it is essential that the marketing of these has been arranged in advance, either privately or through a marketing company, a number of which now operate successfully.

A farmer rearing a batch for beef production on his own farm and a specialist rearer producing calves for sale may differ about their choice of feeding system. The farmer, for example, may opt for minimum labour and an absolute flying start provided by an automatic feeding machine where he is rearing a group of bulls to be intensively fed. The specialist, on the other hand, whilst wanting an attractive group with good bloom at sale, is more likely to put his personal labour into bucket feeding, normally once a day, in order to keep the feed cost per kg of gain to a minimum.

The speed of growth of the calves will also depend on the incoming body-weight, whatever the rearing system chosen. Each extra kg of initial liveweight will be increased to an extra 1.5 kg at least by ten weeks, provided that dry feed is available ad-lib. The extra throughput this could achieve in a unit has to be measured against the relative costs of larger and smaller calves at purchase. Larger calves putting on more weight will also have higher feed costs.

Management Points

If the calves have been carefully selected either by the rearer or his buyer, they will hopefully be free from obvious physical defects. However injuries can occur on the lorry and it is usual for the supplier to guarantee the calves for a week against such injuries and other illnesses. By careful buying it is possible to avoid calves where infections have gained a hold. For example navel infections usually produce hardness and swelling which can be felt. Excessive looseness of the faeces or heavy breathing should also be obvious and indicate infection. Unfortunately even the healthiest looking of calves will have been exposed to infection from other calves, and the stress of a day in the market and the journey by lorry will reduce its resistance. Signs of such illness will usually occur within a few days of arrival as the infection incubates, but symptoms may be delayed until some future stress, such as dehorning or weaning, 'triggers off' a disease. Infections from salmonella organisms can often be delayed in this way.

Once the calves are unloaded and made comfortable in their pens, the most obvious symptom they exhibit is tiredness. Two hours rest without disturbance will revive most of them and they can then be offered something to drink. If they are not keen to drink they are best left until morning, especially if it is well into the evening by this time.

Fairly strong opinions are held on what is the best drink to offer. The most important requirement probably is water to allow them to recover from dehydration which will be greatest in summer. Intake measurement on individual calves at Barhill have shown that they can drink up to 3 litres (5 pints) per day of water in the first three days after arrival.

One of the best and simplest procedures is to offer up to 1.5 litres (2½ pints) of the milk replacer which they will be fed in the rearing period. This should be mixed at the normal recommended rate, e.g. 125 g/litre. This is the type of feed they are most likely to have received prior to arrival and the same as they will be given the next day. A small feed like this in the evening will sustain them through the night. The alternative is to give a solution of glucose mixed at the same osmotic strength as body fluids. For glucose this is a 5 per cent solution or 50 g/litre (1 oz/pint). This will give the calf immediate energy because it is easily absorbed. 1.5 litres of glucose solution will supply 1 MJ of energy at a cost of 7p, compared to 3.5 MJ from the same quantity of milk replacer costing 15p. To put this in perspective, the maintenance requirement of a 45 kg (100 lbs) calf is 8 MJ. If too much glucose is given it can induce a 'sugar scour' so that, on balance, the tendency today is to leave glucose out of the initial programme. A third possibility is to replace the water and soluble salts which the calf has lost during the day. Surprisingly this salty-tasting solution can be quite attractive to calves when used at this time, and it is also a useful therapy if they suffer a bout of scouring later. This provides no energy of course. Various proprietary preparations are available, some of which contain glycine, or a home-made solution can be prepared. Again this must be at the correct osmotic strength. An example of a calf saline solution is shown in table 2.8.

Table 2.8   Calf saline solution

| | To make | |
	1 litre	1 gallon
Bicarbonate of soda	6 g	1 oz
and potassium chloride	6 g	1 oz

If the solution is kept ready made in bulk at double strength, one litre can be diluted with one litre of hot water and the two litres fed at blood heat. After they have received the chosen liquid feed, the calves can be left for the night. A small light may help them to settle; nevertheless some calves will make a determined effort to escape from their pens during the night so it is important to make sure that the doors are securely fastened.

One other practice sometimes carried out is to give the calves vitamins on arrival, either by injection or mixed with the milk replacer. The winter and early spring months would be the more likely time to show a benefit from this practice, because the store of vitamin A in the calf's liver which depends on the nutrition of its dam is likely to be at a lower level in winter. Stated requirements for vitamin A have tended to increase over the years as growth rates have increased, and a better

Plate 11 (*Above*) *Bad practice*—bloat is less likely if the calf is provided with a bucket at head height rather than on the floor

Plate 12 The teat attached by a tube to the bucket allows the calf to suck rather than drink

understanding of the calf's metabolism has been available. However, reputable commercial milk replacers have a very high level (usually 50,000 i.u./kg) of vitamin A included, which will provide well above normal requirements and should make any further supplementation an unnecessary cost. Vitamins C, D, E and the B complex are similarly adequately provided in milk replacers of good quality. The expiry date for vitamin potency is always given on the bag label, and six months or more from the date of manufacture would now be quite typical.

Management over the First Week

The most important objective is to make sure all calves are sucking or drinking properly. Pure Friesians are probably the easiest to teach, whereas Hereford and Continental crosses tend to be more difficult.

For all types of calves, the longer they are left suckling their dams, the more difficult they are to teach to drink from a bucket or suck from a teat. Some calves demonstrate this difficulty for many days, and refuse to drink or suck unless they are 'mothered' by the feeder against whom they like to push bodily. This dam orientation is so strong in 2–3 per cent of calves that they never do learn to drink properly and they wean themselves on to dry feed. It is generally better to let them do this rather than persist with feeding efforts because these cases, and others less difficult, are so agitated and unco-operative when taking liquid feed from the bucket that their mechanism for closing the oesophageal groove does not work. Consequently liquid spills into the rumen instead of by-passing it and going directly to the abomasum. Within a few minutes these calves may develop bloat, the severity of which varies from causing a quick death to mere discomfort. Calves with the latter symptoms often do not eat dry feed well, and grow only slowly. Again they may best be weaned early.

The stockman should have his ear tuned to the cry of a calf in trouble and be prepared to relieve it immediately. The height of the feeding bucket above ground level is important. The optimum height is muzzle level; drinking from pen floor level has been shown to increase the incidence of bloat.

With ad-lib feeding there is less incidence of bloat because the calf normally takes readily to this system. Indeed some rearers on the bucket system put difficult drinkers on to a teat leading from their individual bucket along a short tube. It should not be assumed however that all purchased calves will know how to suck, and each calf in the group must be individually checked and marked with a crayon accordingly.

At the end of the first week up to half the purchased calves are likely to be showing signs of looseness of the faeces. As long as this is not

very watery or accompanied by straining or by blood, the condition can be put down to the change and build-up of diet. It should not be classed as scouring and no action is needed unless it is felt that a slight reduction in the quantity of milk replacer would help. Note that the strength of the mixture should not be changed. Calves on the ad-lib system show the most marked excessive faeces production between one and two weeks after arrival because of their high intakes of milk replacer of about 1 kg per day at this stage.

Pre-Weaning

By three weeks from purchase the calves should be looking much cleaner, with firmer faeces, even though they will still only be eating quite small amounts of dry feed. General management of the purchased calves will be the same as home-bred calves, except for extra vigilance to detect the development of disease symptoms. This applies particularly to pneumonia and to several other diseases which are dealt with in Chapter 5. As a matter of routine all the calves should be dusted with louse powder which will prevent the build-up of lice before the calves start to rub their hair off. Remember that the objective is to rear an attractive animal for sale. Similarly signs of ringworm are best treated early to give more time for the regrowth of hair before sale. Finally the important job of disbudding the calves is best done early, since the horn buds develop rapidly from three weeks of age. In planning the work routine, it should be noted that since January 1983 calves over eight weeks of age can be castrated only by a veterinary surgeon using an anaesthetic.

Post-Weaning

When the calves are weaned at about five weeks after purchase they will need more room than their initial pen provides and they can be grouped in pens of eight to ten. Bigger groups can be managed but take more observation time as well as careful siting and allocation of feeding space. Thinking ahead to the time of sale, selection of calves either by eye or over the scales to form uniform batches will provide more attractive groups, and reduce competition.

Feeding after weaning is very much simpler than before as long as the early-weaning concentrate is kept available on a truly ad-lib basis. Attempts to ration the calves because it is thought that they will produce cheaper liveweight gains or because over-consumption will be prevented are counter-productive. Of course a good design of hopper or trough is needed to prevent waste, but calves will not over-eat or upset themselves when the feed is always available. On the contrary it is attempts to gauge appetite and to put feed into an empty trough that lead to bullying, over-consumption by some calves followed by scouring or bloat.

As previously stated a feed conversion of 2.8:1 between five and twelve weeks means that the liveweight gains will be costing about 60p/kg against a sale price of 70p. This is cheaper than the cost per kg pre-weaning, and cheaper than it will ever be after twelve weeks.

Clean water must be easily accessible to the weaned calves which will now be drinking in the region of 10 litres (2½ gallons) per day. For young calves self-filling bowls are better than treadle bowls. They should be checked daily to ensure they are not fouled or frozen up during cold weather.

If all goes well the calves should average 100 kg (220 lbs) body weight at eleven weeks from an average start weight of 45 kg (100 lbs). Progress can be assessed by weighing a number of calves or by using a weigh tape pulled tightly round the chest just behind the front legs.

Buying and Selling Calves

Sale at around twelve weeks is best based on a bulk weighing of calves on the lorry by public weighbridge, where a duplicate ticket of the net weight of the calves can be obtained.

Overcrowding on the lorry must be resisted if losses in transit are to be avoided. Partitioning gates should separate the calves into small groups of eight or ten. The heat of the middle of the day should be avoided in summer if at all possible. The rearing of calves for sale to beef producers is carried on as a marketing exercise by many co-operative groups, commercial companies and by individual farmers and dealers. It is essential for the rearer to have established his marketing arrangements before starting up an enterprise. Sales are best direct from rearer to fattener who is looking for an established group of calves which are past their early health problems. Commonly these batches are required in groups ranging from twenty to a hundred. The rearer cannot fall back on being able to sell these reared calves in the open market. At this stage, intermediate between a young calf and a true store animal which can be turned out to grass, reared calves never realise their true worth at auction, even making less than the original cost price of the calf.

The advantage of using a co-operative marketing organisation is that it encourages a better discipline all round. Although there has to be a collection centre (lairage) and mixing of calves, there is less stress than in open markets and only one move from breeder to rearer. With the staff of the co-operative dealing personally with both buyer and seller and keeping a record of calf identification numbers, there is every incentive to present calves well.

Where calves are being marketed by a company or group, the final purchaser can be informed of the price to him at the time they put into the rearer. He may ask to see them either during the rearing period or just prior to sale. The most common area of disagreement at sale is over the type or conformation of the calf, especially for pure Friesian calves which will often be going into an intensive beef system. Over recent years this problem has been aggravated by the increasing use of Holstein bulls which stamp their offspring with large angular frames— the opposite of a compact beefy animal. The proportion of Holstein blood in a calf can of course vary anywhere from nil to a hundred per cent. The calf buyer can only use his experience and judgement in trying to select calves which will turn out to have good beefing qualities. The best of these men will tell you that this cannot be done with absolute certainty. Add to this a fair degree of prejudice about colour, with white calves being generally disliked, and it can be seen that it is no easy task in getting a group of rearing calves together at a reasonable price.

Although the Friesian bull calf is most affected by the Holstein influence, beef crosses out of Holstein cows can also show poor conformation. Of these, Hereford crosses are the most numerous and suffer from having a poor thigh which can be sufficient to have them downgraded also. Since 1987 the Continental crosses, Limousin, Charolais, Belgian Blue and others dominate in the markets. So the rearer must be vigilant in the type of calves he accepts for rearing, especially as one or two calves of poor type in a group tend to down-grade the other calves by association, however unfair that may be.

Making Up a Group
Whatever the eventual size of a group of calves for rearing, the ideal is to purchase them all over a short period of time, and not mix them with other calves thereafter. This will help them to build up their common resistance to infections which they brought in, without further challenge from late arrivals. Although this throws more strain on the feeder for the first few days, it also eases the planning of feed deliveries, work routines and delivery dates to customers. When sale time comes uniformity is important. However well reared, the range in weights of the incoming calves will have widened three months later at the time of sale. It is therefore more satisfactory to be able to split a large batch into two or three smaller uniform batches. Records from a very large number of calves reared at Barhill show that every 1 kg difference in

liveweight at purchase will, on average, widen to 1.3 kg at five weeks and nearly 1.5 kg by ten weeks. An example (see table 2.9) will make this clearer and shows that the extra gain of the heavier calf comes from a higher dry feed consumption.

Table 2.9 Expected gains from two initial liveweights

Initial liveweight		Liveweight gain		Dry feed consumption		Feed conversion milk replacer and dry feed
(kg)	(lbs)	(kg)	(lbs)	(kg)	(lbs)	
Five weeks from purchase						
38	(84)	16.7	36.8	12.5	27.6	1.51:1
50	(110)	19.0	41.9	22.0	48.5	1.82:1
Ten weeks from purchase						
38	(84)	45.9	101	87.8	193	2.19:1
50	(110)	50.4	111	113	248	2.42:1

The 50 kg calf will have reached 100 kg at ten weeks and be ready for sale. The smaller 38 kg calf has been more efficient in feed conversion over the ten weeks but has only reached 84 kg. Its cost per kg gain was lower at 82p as against 85p for the larger calf, largely because its energy for maintenance has been less for its lower average weight (10 MJ as against 11.5 MJ per day) over the rearing period.

However, the smaller calf still has to put on 16 kg before it reaches the saleable weight of 100 kg so that its total feed costs will be greater, as shown in table 2.10. Also the extra time taken of about eighteen days means that throughput in the unit will be 26 per cent less and that overhead costs, particularly of labour, will be greater. With a minimum cleaning and rest time of fourteen days between batches, purchase of the smaller calves would achieve 3.5 batches and the larger calves 4.3 batches per annum.

Table 2.10 Feed costs to selling weight

38 kg calf at purchase	£	50 kg calf at purchase	£
12.5 kg milk replacer	18.75	12.5 kg milk replacer	18.75
88 kg dry feed	18.43	113 kg dry feed	23.73
50 kg dry feed for a further 16 kg LWG	10.50		
Total feed cost to sale	47.68		42.48

The difference in purchase price of the lighter calf would have to be considerably more than the difference of £5.27 in feed cost to make it a more attractive proposition, bearing in mind the slower throughput and lower resilience of the small calf.

Sale of Calves
As the time of selling approaches it is a good idea for the fieldsman of the selling organisation to see the calves available and for him to bring the customer with him or keep him informed of his visit.

The rearer needs to know batch numbers required and can then make up appropriate groups, weeding out any not fit for sale. Reasons for this may be that:

- a calf is too small even though it is thriving;
- it may have developed into a very poor type;
- it may have become ill or injured.

Obviously the aim is to keep unsatisfactory animals to a minimum because they will make an overall loss if they are disposed of at market value. On the other hand the customer, if well satisfied, is likely to become a regular buyer. Most farmers would rather take one or two calves less than ordered, if this has resulted from leaving out the doubtfuls.

An early decision on whether the calves are to be castrated enables full recovery before sale and minimises the cost to the rearer because of the temporary check in growth.

Delays caused by late castration or any other reason work against the rearer. He has nothing to gain and a lot to lose from the costs and risks of having the calves on the rearing unit when they are ready to go. Injury or illness incurred is often a severe loss with a marked effect on the net margin per animal in the group.

Feed costs also accumulate very rapidly near the sale stage since it is likely that the calves have remained on ad-lib feed. On a straight conversion basis the feed cost will be paid for by the increase in liveweight gain, but there is no extra income to cover labour and other overheads, or the reduced income from the delay in taking in a new batch.

Seasons of Rearing

Where facilities and labour are committed to a calf-rearing enterprise it is clear that these must be employed as near to capacity as possible by rearing all the year round. However, if other circumstances allow a more flexible approach the most advantageous seasons can be chosen.

Reared calves for grazing are normally in demand, either beef crosses or to a less extent Friesian steers, but they need to reach 180 kg (400 lbs) bodyweight minimum at turnout. Their start date will therefore have to be during October with latitude of one month on either side. Since this time coincides with the greatest supply of calves from the dairy herd, and usually the lowest prices in the year, this fits in well and is a good time to rear. Additionally, such calves are born at the end of the summer and are usually healthy and well developed.

Therefore if follow-on accommodation is available, one or two batches of say, Hereford cross Friesian steers could be taken through the winter and they would find a ready sale as soon as spring grass is available. Some purchasers of this type of stock often make private arrangements to have their calves reared. This has the advantage to the rearer and the purchaser of a known price in advance. Also vaccination for husk can be requested, paid for and carried out in the six weeks before turnout time.

In the spring there is another flush of calves born during February to April when again it is easier to obtain the type of animal required. From then on numbers dwindle and prices generally harden. Summer-born calves have the further disadvantage that they are ready for sale in the autumn when market prices of young calves are low, making the price for three-month-old reared calves look very expensive. However, the pricing of milk each month is now aimed at producing a level supply and thus to a more level supply of calves.

Early spring-born calves are reputed to have low vitality because of the poorer nutrition of their dams during the winter. Although this is probably less true today than in times past, nevertheless there is still some truth in it and some calves will certainly be less able to resist infections with which they come into contact if weather conditions are adverse. However, examination of results from thirty winter-reared lots of calves and thirty summer-reared lots at Barhill showed no difference in performance to five weeks. Both groups had liveweight gains of 18 kg (40 lbs) which was an average for different bucket-feeding systems. Post weaning there was a slight advantage to the summer-reared calves of 1.5 kg (3½ lbs) liveweight gain for the 5–10 week period. This extra gain arose from a higher intake of 4.5 kg (10 lbs) of early-weaning concentrate so that feed conversion efficiency remained the same. For the purpose of this comparison winter was defined as the period from November to April.

So the difference is likely to be relatively small and overridden by standards of housing and management on individual farms. An interesting comparison has also been made using an acidified milk replacer compared to an equivalent non-acidified one. Both products contained over 60 per cent skim milk powder and were fed in the normal bucket system. On two occasions the acidified powder gave a liveweight gain advantage when the trial was run during the late winter but showed no advantage when the same diets were compared in summer. The advantage to the acidified diet was 1.8 kg for one trial and 2.5 kg for another for the period of 0–10 weeks. Dry-feed intakes were equalised in both trials. Both results were statistically significant in favour of the acidified milk replacer, and are not likely to have been chance effects.

Although no scientific explanation can be given for a better performance on acidified diets, quite a number of rearers have confirmed this advantage by observation, and use such diets either in winter only or all the year round. The use of acidified products for replacement calf rearing on the bucket is almost universal in Holland, whereas in this country they were primarily developed for and are used in the cold ad-lib system. Part of the explanation for good performance on the ad-lib cold system may lie in the acidification of the diets, but it has not been possible so far to separate this factor from the benefits associated with higher intakes, frequent feeding and extra salivation which also occur with this system.

Choice of Breed

The types of calf reared in the largest numbers are Friesian, Hereford cross Friesian and Continental bulls cross Friesian. Of these the pure Friesian bull calf is easiest to rear. It is quick to learn to drink except for a very few individuals, whereas up to 20 per cent of the Hereford cross bulls can be difficult drinkers. Continental crosses such as the Charolais or Limousin can be more difficult still. Most rearing companies pay an extra fee for beef crosses to cover the higher interest charge on the purchase price, usually 50 per cent higher than for pure Friesians. The cost of any mortality is also 50 per cent higher and these together amount to £6 at 1995 prices, to say nothing about frayed nerves. During 1977, in spite of very high purchase prices, rearers tended to favour rearing beef crosses, which were more attractive to the purchaser partly to minimise the risk of losing variable premium.

An investment in Hereford cross Friesian heifers would cost about the same as for Friesian bulls. However they can be just as difficult to rear in the early stages as Hereford cross bulls but, at least, the task of castrating before sale is avoided.

The performance of the heifers will not be quite up to that of the bulls on average. Records show that 576 Hereford cross heifers reared at Barhill gained 2 kg less to five weeks than Friesian bulls over the same six-year period on various bucket systems. At ten weeks the liveweight gain was 6 kg less and although dry feed intake was also 8 kg less, overall efficiency was 6 per cent worse with the Hereford cross heifers. As the beef men know well, beef cross heifers lay down fat much earlier than steers and even in the calf stage there is the same extra energy requirement for growth. However, there is a demand for these smaller carcasses provided they are not allowed to become over-fat. The Barhill system involves buying in October and selling the following October–December at 200 kg deadweight off grass.

The rearer is to a large extent not able to pick and choose his breed to rear—he has to meet the demand as it exists but can make his preferences known.

Chapter 3

CALF PHYSIOLOGY AND NUTRITION

CALF DIGESTION

ALTHOUGH THIS is not a scientific textbook it is important for the calf rearer to understand the basic concepts of how the calf digests its food and how it obtains its main nutritional requirements.

Although destined to become a ruminant animal with four functioning stomachs which give it the ability to deal with fibrous feeds, the calf at birth has not developed this capacity. The newborn calf's true stomach, or abomasum, is the only functional stomach in the early stages. Milk or milk replacer, whether it is sucked from a teat, or drunk from a bucket is channelled directly from the oesophagus (gullet) via the 'oesophageal groove' which forms for this purpose in response to various stimuli. As mentioned in the management chapter, this device appears to work perfectly when the calf is suckling its dam and persists for several months with the suckled beef calf. In a small proportion of artificially reared calves the mechanism fails to respond to the stimulus of milk or milk replacer from the bucket and this appears to be a psychological condition in response to being separated from its dam.

Most calves can be trained by patient coaxing to drink quickly and well, responding to the new daily routine and substitute dam in the shape of the calf rearer. Interestingly this sometimes involves being in the pen with the calf and allowing it to push against the feeder as it would against its dam. When milk or milk replacer with a high proportion of skim milk powder enters the abomasum it forms a firm clot within a few minutes under the influence of the enzymes rennin and pepsin. This is the same process used by cheese makers with rennet which causes the milk protein to coagulate. Similarly rennet is used for making junket.

Any fragments from the previous feed will be enveloped in the newly formed clot. There is then a rapid separation of liquid whey protein and lactose (milk sugar) which pass into the duodenum (small intestine) at a rate of about 200 ml (7 fluid ounces) per hour.

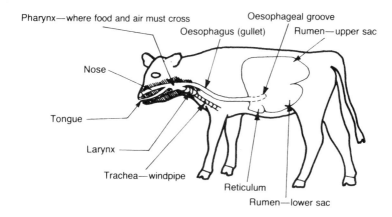

Pharynx—where food and air must cross
Oesophageal groove
Oesophagus (gullet)
Rumen—upper sac
Nose
Tongue
Larynx
Trachea—windpipe
Reticulum
Rumen—lower sac

Figure 3.1   The upper digestive tract of the calf. The rumen would be proportionally smaller than this in a young milk-fed calf

Within the protein curd is embedded the milk fat and it is believed that partial digestion of this fat is accomplished by an enzyme, lipase. This is secreted in saliva and is incorporated as the milk is swallowed. More saliva is produced when a calf sucks from a teat, rather than when drinking quickly from a bucket and this could be an advantage for the ad-lib systems. This fat digestion at the pre-gastric stage is more efficient with butterfat than with other fats used in manufacturing milk-replacer powders.

Of the two enzymes acting on the milk protein, rennin is more effective than pepsin, particularly at the near neutral pH which occurs in the stomach immediately after a feed. The proportion of the two enzymes varies a good deal from calf to calf. Within a few hours the acidity in the stomach increases, until just before the next feed it is extremely acid (pH 2.0) from the influence of hydrochloric acid secretion.

With milk replacers based on a high level of skim milk powder a good firm clot is formed in the abomasum as it is with whole milk. However if in the drying process at the creamery too high a temperature is used, the resultant overheated skim milk powder will exhibit a poor clotting ability. Moreover the protein will be downgraded in value and, if severe overheating has taken place, digestive disturbance and a greater risk of illness can be caused.

Severe overheating and caramelisation are more liable to occur with roller-dried powders, but these should not be found in high-quality powders which are mainly used today. Nevertheless, even with the

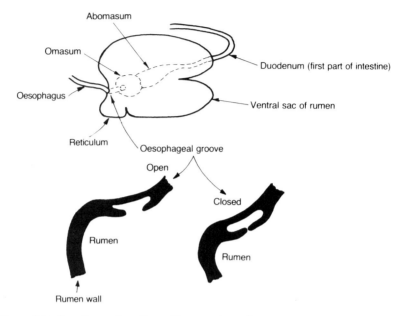

Figure 3.2   Function and position of the oesophageal groove

spray drying process, degrees of over-heating can occur and must be monitored by an efficient quality-control system. The new generation of milk replacers, so-called 'zero' diets because skim milk powder is absent or at a low level, are based on whey products of varying types and may have additional protein from the inclusion of processed soya or fish, or from single-cell protein. These milk replacers do not form a clot in the abomasum and this lack of clotting ability can be demonstrated in the laboratory. Because such milk replacers can produce entirely satisfactory results in calf rearing, it must be accepted that the calf is able to adjust to this non-clotting feed. For good results it is essential that the 'zero' formulations are based on good test work and that good quality control of raw material applies.

Further digestion of the protein and fat takes place in the small intestine with the aid of enzymes produced by the pancreas. Lactose which is released quickly from the clot in the abomasum is further broken by the enzyme lactase into glucose and galactose which are absorbed quickly and provide immediate energy to the calf. The young calf is not able to digest starch in the first weeks of life, but a small amount of high-grade starch is included in all milk replacers based on skim milk powder to 'denature' it within EEC regulations. Proteins are broken down into constituent amino acids, and fats into fatty acids and glycerol after being emulsified with the aid of bile salts from the gall bladder. Digestion in the small intestine takes place under alkaline conditions which result from the pancreatic juice inflow.

It can be seen that the calf maintains a complete acid and alkaline status in the digestive system. Any nutritional disturbance or bacterial infection causing scouring not only dehydrates the calf but results in the loss of mineral salts or electrolytes.

Rumen Development

The first stomach or rumen develops from a very small initial size with only about 2 litres capacity. On the early-weaning system it enlarges quickly during the first weeks of life. However development of the rumen wall only occurs under the influence of volatile fatty acids produced from the fermentation of dry feed. The rumen volume and dry feed intake are positively related. Development of the rumen wall in a veal calf is almost nil when it is deprived of dry feed or roughage. By contrast, on the early-weaning system where milk replacer is fed in restricted quantities to encourage dry feed intake, the rumen size and the papillae, which are leaf-like structures on the internal surface, develop in the early weeks. Cudding is initiated at about two weeks of age.

With ad-lib systems of rearing there is no incentive for the calf to take dry feed though it is freely available. The calf satisfies its appetite with its liquid intake and it needs a period of restricted availability of liquid feed to stimulate dry feed intake.

The rumen papillae increase its surface area many times and greatly enhance the absorptive capacity for the end products of digestion. These are mainly the volatile fatty acids, acetic, propionic and butyric acids. The last two of these in particular stimulate the development of the rumen and come mainly from the digestion of the dry concentrate diet. However, the presence of fibrous material either mixed in the concentrate or available on free access is beneficial and also helps to maintain the optimum pH in the rumen. The degree of acidity is known to affect the type and efficiency of the papillae.

On the other hand the view that good-quality 'fine' hay eaten in relatively greater amounts than dry concentrates will stimulate optimum rumen development is now outdated. The rumen is certainly stretched to accommodate the mass of fibrous material and this gives rise to a typical pot-bellied appearance which is undesirable. The best results for good rumen development and growth come from an ad-lib supply of specially formulated concentrate feed, together with hay or preferably clean barley straw on offer until ten to twelve weeks of age.

Digestion in the rumen takes place entirely by fermentation of the ingested feed by millions of bacteria and protozoa which become naturally established early in life. The relatively larger protozoa, still microscopic in size, increase slowly in numbers as the rumen pH comes nearer to neutrality and with increasing roughage. The teeming population of micro-organisms in the rumen establishes itself quite naturally. Occasionally, if the calf has been treated by mouth with antibiotics, it may be necessary to reinoculate the rumen with the cud from another calf.

As would be expected from a fermentation process, carbon dioxide and methane gases are produced and exhaled. When something prevents the escape of these gases bloat can result at any stage in a ruminant animal's life. In the young calf the commonest time for bloat to occur is a few minutes after a liquid feed. This is often due to milk replacer spilling into the rumen and being rapidly fermented. In four successive years from 1978 to 1982 at Barhill, annual losses from this type of bloat have been three, two, one and three respectively out of twelve hundred calves reared each year.

Occasional 'outbreaks' of three or four calves dying from bloat on a single farm are reported and it is difficult to account for them. More trouble is generally experienced with beef cross calves and Channel Island calves than with Friesians.

NUTRITIONAL REQUIREMENTS OF THE CALF

Energy

Energy is needed to maintain body temperature and to support the calf's body functions. This is known as the maintenance requirement. Any energy taken in by the calf which is surplus to this basic need is available for growth, the laying down of muscle and fat which is generally called liveweight gain. Energy requirements and the available energy in feeding stuffs are now measured in units of the Joule replacing the more familiar calorie (1 calorie = 4.184 Joules). Most farmers are now conversant, for example, with the energy value of average grass silage at 10 mega (million) joules (MJ) per kg of dry matter or with the requirement of 5.2 MJ from feed per litre of milk of average compositional quality.

Similarly the total energy required to put on liveweight gain can be expressed as MJ per kg of gain. This figure when quoted would normally cover the requirement for maintenance as well as that for growth. Maintenance requirements increase in direct proportion to age and size and are expressed scientifically as the need per unit of metabolic liveweight (kg $LW^{0.75}$).

However, with growth, the energy requirement per kg of gain not only increases with age and size, but it also varies with the energy concentration of the diet. High-energy diets are considerably more efficient than low-energy ones for growth. In other words, more megajoules per kg of liveweight gain are needed on a low-energy feed compared to a high-energy feed. Figure 3.3 should help to make this clear. The bottom line denotes the energy in MJ for maintenance, whilst the two upper lines show the total energy in MJ required for 0.5 kg/day and 1.0 kg per day of liveweight gain. For example, at 100 kg liveweight the calf requires a total of 50 MJ/kg gain when growing at 0.5 kg per day compared to 34 MJ/kg when growing at 1 kg per day. Since the animal's appetite for dry matter (DM) is more or less constant at any given liveweight (it is about 100 g DM per kg of metabolic liveweight) it follows that the energy concentration of the total diet for the higher rate of gain must also be higher. This is expressed as Megajoules over dry matter (M/D) and overall diets normally lie in the range 8.0 to 12.0 with individual components from 6.0 to 18.0 MJ/kg DM.

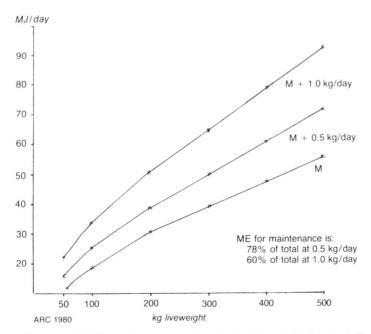

Figure 3.3 Energy required for maintenance and two levels of growth for Friesian bulls

Figure 3.3 shows clearly that the provision of the maintenance part of the diet is a greater percentage of the total energy for 0.5 kg/day compared with 1.0 kg/day.

Also as the calf grows and its liveweight increases the proportion of energy used for body maintenance becomes greater. At various liveweights through to the adult stage, the proportion of total energy used for maintenance purposes is shown in table 3.1.

Table 3.1   Percentage of total energy (M J) used for maintenance at various liveweights

LW (kg)	1.0 kg/day (%)	0.5 kg/day (%)
50	30	36
100	45	59
300	51	68
500	55	71

The overall average, from calfhood to slaughter at 500 kg, for the two rates of gain shown are 47 per cent and 61 per cent respectively. Because, by definition, the time to reach 500 kg is twice as long for the slower-growing animal in our example, the total energy required can be calculated at 1½ times as much.

The example shown has widely spaced standards of performance though these are not outside normal practical limits. Indeed, most progressive eighteen-month beef producers would aim at 0.85 kg/day at grass and 1.2 kg/day in the final winter. The example has also gone beyond the calf stage right through to slaughter in order to demonstrate the progressive deterioration in energy efficiency between 50 and 100 kg LW and onwards.

The change in efficiency can also be stated in terms of feed conversion ratio (FCR) which shows the same picture (table 3.2).

The lesson to be drawn from these data is the same for the rearer who is selling at 100 kg LW or for the finisher who takes reared calves through to slaughter. It is that early gains are the most economical in energy terms and the high efficiency in early life should be exploited to the full.

Table 3.2 Feed conversion ratios at various liveweights

LW (kg)	FCR (kg gain per kg DM intake)
50	2:1
100	3:1
300	5.5:1
500	8.5:1

The cost per kg of gain depends not only on the efficiency of gain, of course, but also on the cost of the energy. The most expensive period in the calf's life is during the first five weeks when fed on milk replacer as well as dry feed. For bucket feeding once a day, the cost on average is about 135p/kg gain over this period, and for machine rearing the cost rises to 210p/kg.

Fortunately this high cost is only for a short period of time and is an essential part of getting the calves off to a good start. It cannot be sensibly avoided. After weaning, from five to twelve weeks, the most economical period occurs provided that good-quality dry feed is given ad-lib. Cost per kg of gain will be reduced to about 60p.

The reader will forgive re-emphasis of the above point because one of the commonest mistakes made by rearers is in not making feed available on a truly ad-lib basis at this stage, or in reducing protein level too soon. By and large contract rearers appreciate the necessity for high economical gain up to twelve weeks better than rearers of dairy heifers.

From 100 kg liveweight onwards the cost per kg of gain will go on rising and can reach £1.20/kg in the later stages of beef finishing which is as much as the beef is worth. It will largely depend at this stage on the total amounts and quality (digestibility) of the concentrate feed, grass and roughages which are available, and the unit price of energy in each of them.

In Chapter 2 reference was made to the differences in efficiency of feed use between bulls, steers and heifers. Taking the energy requirements for growth for steers as the standard, the energy efficiency for bulls would be 15 per cent better, and for heifers 15 per cent worse. Such differences in energetic efficiency are more important economically in the finishing period rather than in the early calf stage because of the greater absolute amounts of daily energy needed in the later stages.

The exact requirements of these different types of stock are now

accurately stated in the 1980 ARC publication *The Nutrient Requirements of Ruminant Livestock*, which most advisers now use for the
formulation of rations. Energetic efficiency can also be positively
influenced by growth promoters which many compounders add to their
calf feeds. Examples are nitrovin, bambermycin and avoparcin. The
inclusion level of these growth promoters must be declared on the bag
or ticket label of compound feeds including milk replacers, together
with the product licence number.

The growth promoters act on the bacterial flora of the rumen in
particular but also in the small intestine, promoting a more efficient digestion. For the maximum effect they should be included at
the correct level for liveweight throughout the rearing period up to
turnout.

Readers will be well aware that since January 1987 (a year later in
the rest of the EEC) the much more potent implants of anabolic agents
have been banned. This ban has been on both natural and synthetic
compounds and has had the effect of slowing down growth rates in
both steers and heifers.

Protein

Proteins are required by the calf to maintain biological processes on a
daily basis, as well as repairing tissues and forming blood. Proteins are
also an integral part of muscle growth, predominating in the laying-
down of lean flesh. All proteins consist of complex compounds containing large molecules made up of over twenty 'building blocks'
called amino acids. Some of these are well known by name, more
especially in pig and poultry nutrition where the diet has to supply the
complete requirements of essential amino acids. Of these lysine is
probably the best known and is available in synthetic form as a dietary
supplement.

Another amino acid, methionine, is thought to be the first limiting
one in ruminant diets. This can also be synthesised and is often added
to calf milk replacer diets containing substitutes for milk protein such
as processed soya. Responses to formulations where deficiencies of
specific amino acids are rectified are easy to demonstrate with pigs and
poultry.

By contrast, however, with calves and adult ruminant animals, little
information is available on the precise needs for specific amino acids in
the diet. In the early stages the dietary make up can be likened to the
amino acid profile of cow's milk. Once rumination is active the amino

acid needs of the ruminant animal are partly provided by synthesis in the rumen of microbial protein which is composed of 80 per cent amino acids, the profile of which is believed to be good. Other amino acids are provided by undegradable protein in the diet. Undoubtedly limiting amino acids will be discovered in the high-performing ruminant. Work is likely to be concentrated on the high-yielding dairy cow in the first instance because of her high nutritional needs to provide a high output of protein in milk. It is also being actively studied in the diet of young calves up to the twelve-week stage.

The element nitrogen is an essential constituent of all proteins, present at about 16 per cent, though varying slightly with different proteins. This is why when feedstuffs are analysed for protein, the chemical test determines the nitrogen level which is multiplied by 6.25 ($\frac{100}{16}$) to give the protein level, or 6.38 where milk products are concerned. The use of the nitrogen level to determine protein by the chemist is similar to effects of the teeming microbial population in the rumen. These bacteria and protozoa are able to use simple nitrogen sources, principally ammonia, to form their own cell protein as they grow and multiply.

The ammonia used by the microbes comes from the breakdown of proteins and other simpler nitrogen compounds in the feed eaten, but it can also come from simple salts such as urea. This can be manufactured synthetically and is sometimes added in limited amounts in the diet of older ruminants (up to 1 per cent of the total diet). Urea may also be used as a nitrogenous fertiliser, where it is first turned into plant protein before eventually being made available to the animal as plant feed such as grass or silage.

The bacteria which have grown in the rumen are in due course passed on down the alimentary tract and are digested in the same way as normal protein, providing amino acids. In this form they are absorbed and utilised within the body system.

The fermentation process, i.e. the building up of the microbial biomass, needs a supply of energy provided by the breakdown of sugars, starches and fibre in the feed. Both processes take place concurrently with different organisms which exist in many millions per millilitre. The supply of energy sets the limit on the growth of the microbes and thereby on the 'microbial yield' of protein. Whilst microbial protein synthesised in the rumen is by itself sufficient for low levels of performance, it will not support fast-growing animals or, more particularly, high-yielding dairy cows which require undegradable protein supplied additionally in the feed.

This realisation has brought about a new dietary protein concept of supplying a sufficient nitrogen source in the rumen available to the microbes which is known as degradable protein (RDP). Other protein in the feed, which because of its more resistant make-up escapes breakdown in the rumen, is called undegradable protein (UDP). This is largely digested further down the digestive tract, in the abomasum and small intestine.

Raw materials for inclusion in diet formulation can now be selected and blended to supply the correct balance of both types of protein. In practice there is usually adequate RDP supplied from farm feeds and many concentrated feeds. Where extra UDP is required within the total protein of a dairy or calf compound, for example, then constituent proteins have to be selected for their relatively high level of UDP. Fishmeal is perhaps the best known of these but it is expensive and in limited supply; other good examples are dried distillery products. Soya bean and rapeseed meal are major sources of protein, but because they are fairly highly degradable, processing has been developed to render their protein more resistant to attack in the rumen.

The processing may be simply a carefully controlled heat treatment or the use of formaldehyde, which bonds chemically with the protein. Both processes add to the cost of the raw material and have to be valued carefully against unprocessed protein feeds simply included at a higher level.

The final twist to the story is that if the proteins are over-protected, either naturally as can occur with tannins, or artificially by chemical or physical treatment, then they will be digested to a very limited extent. This can easily happen for example where skim milk powders are over-heated when drying, thus reducing digestibility in the calf's stomach.

How does all this affect the calf? In the first few weeks of life the production of microbial protein in the rumen is not an important protein source to the calf, provided that an ample supply of milk is being directed into the abomasum. This will be the case when whole milk or milk replacers containing high-quality skim milk powders are being fed. These have a digestibility of over 90 per cent, all of the protein counting as UDP because it bypasses the rumen via the oeso-phageal groove.

Under the early-weaning system, with bucket feeding, the amount of such milk protein provided is limited as part of the plan to encourage the calf to eat dry feed in sufficient amounts to wean without a check in performance. This dry feed has to stimulate the rumen to manufacture microbial protein, and also provide a good supply of undegradable protein needed in addition. The critical time for protein supply is obviously when the milk protein is withdrawn at weaning from the bucket. From then on the calf is dependent on the dry feed for protein and, over the next four or five weeks, the supply of both RDP and UDP will increase slowly as the intake of dry feed gains momentum.

The development of the rumen is indicated to the stockman by the calf chewing its cud which can commence as early as two weeks of age.

For the two weeks before weaning and for the three or four weeks afterwards, maximum growth of 0.7 to 0.8 kg per day for the early-weaned calf can only be achieved if a high-protein dry feed is being eaten by the calf on an ad-lib basis. Such a feed would probably have an 18 per cent crude protein content with about a third of this in an undegradable form. It also needs to be highly palatable and attractive to the calf, and this is more important than its appeal to the farmer.

It is instructive to look at the daily intakes of dry feed on a week-by-week basis through to twelve weeks. The example on Graph 2 in the Appendix shows the average intake of a large number of calves fed on the once-a-day bucket system at Barhill. These produce a sigmoid curve with a large increase in daily intake between weeks five and six, from 1.3 kg to 2.1 kg per day which is a 60 per cent increase. Once the intake reaches about 3 kg per day the contribution of microbial protein will be sufficient to allow a reduction to 16 per cent protein in the feed. Later reductions to 14 per cent and eventually to 12 per cent crude protein will depend on the level of feeding practised and also on the protein supply from other feed sources such as silage.

Turning to the ad-lib system of feeding with the young calf, the graph also shows the intake of dry concentrate feed for such calves which received around 30 kg of milk replacer by the time they were weaned at five weeks. Their dry feed intake is much reduced, reaching only 0.64 kg per day during the fifth week after negligible intakes during the first three weeks. However there is a massive increase of 2½ times in intake during the week after weaning when it can reach 2.1 kg per day.

If there is no incentive to start eating dry feed in weeks four and five by a restriction of the supply of milk replacer from either the containers holding the cold acidified products, or warm from a machine, then sudden weaning from liquid feeding will produce a massive deficit of both energy and protein through insufficient intake of dry feed. This will check the calf's growth rate very considerably.

The data on intake from these two methods of rearing have been standardised at five weeks weaning for comparative purposes, but it is appreciated that individual rearers may wish to adopt their own variations in weaning techniques and in timing aimed at smoothing the changeover. Those rearers using high milk replacer intakes on machine feeding are the ones who most need to assess their best weaning procedures. It would also be easiest for them since the extension of the period of restricted access only involves switching the machine on and off each day, and probably also the replacement and removal of the screw-on teats.

Crude Fibre

This term conventionally covers a complex of substances which are digested by the action of micro-organisms, mainly cellulose and hemi-cellulose, which may be bonded with woody tissues (lignin) to a variable extent. The younger a plant, the less woody is its fibre fraction and the more digestible it is. The concept of digestibility or 'D' value (digestibility of the organic matter in the dry matter) of grasses at the time of silage or hay-making is well known. It is used to denote the stage of growth of the grasses and their likely feed value. The calf does not have a nutritional requirement as such for fibre in the sense used for energy and protein requirement. Indeed fibre is made available by microbial fermentation in the rumen.

Veal calves are fed on a diet which is almost fibre free, but calves which ruminate do need a proportion of long fibre in their diet. However the real contribution of fibre to the calf diet does not come until it has reached about 100 kg liveweight when good-quality hay, straw or silage can form an increasingly important part of the total diet.

Fibre does make some contribution to the energy requirements of the weaned calf and, more importantly, aids the development of the rumen. This part of its function is involved with increase in rumen size, absorptive capacity and the maintenance of an optimum pH level. Additionally the salivary glands are stimulated by fibre intake.

The calf can take in fibre from hay or straw in long form. Also there will be a proportion of fibre in the dry concentrates, especially from the inclusion of materials like oats, dried brewer's grains, sugar beet pulp, nutritionally improved straw or dried grass which are all relatively high in fibre content.

Adequate fibre in the diet helps to slow down the rate of fermentation in the rumen and to prevent excessive acid conditions (below pH 5.5) which are known to reduce cellulose digestion, impair intake, reduce turnover rate and slow down growth rate.

The total fibre in a compounded calf ration will be stated in the guaranteed analysis, together with protein, oil and ash. Vitamins are also declared. Unfortunately the crude fibre declaration gives no indication of fibre digestibility and thereby value to the calf. Good-quality compounds would normally be expected to include fibre sources which are highly digestible to the calf.

Another approach to keeping the pH of the rumen above a reading of 5.5 has been to include buffer salts such as sodium bicarbonate, at about 2 per cent of the diet. Although a more common practice in dairy cow feeding, the bicarbonate has proved effective in diets which were high in rapidly fermentable starch and sugars, especially where cereals were processed by cooking or micronising. However sodium bicarbonate is expensive and will dilute the energy levels according to its inclusion rate unless special steps are taken to balance it. Inclusion of digestible fibre is the preferred course of action.

Minerals

Calcium and phosphorus are the two major minerals required for building the calf's skeleton which has a high content of both of them. A deficiency of these two minerals is associated with rickets, and phosphorus deficiency is also associated with reduced appetite or a depraved appetite. Vitamin D is also required for proper calcification of the bones and rickets can occur if vitamin D is deficient, even though adequate calcium and phosphorus are present in the diet. The other major minerals—magnesium, sodium, potassium and chlorine—are all required in the diet and can be found mainly circulating in the blood and soft tissue of the calf. These minerals are the ones which become deficient in scouring calves and need to be replaced by electrolyte therapy.

Calf rearers will generally assume that the milk replacer and dry feeds which they buy will contain the correct level of minerals. Because this is usually the case it is rare for problems to arise because of mineral deficiency or imbalance, but as higher standards of performance are achieved it is possible that certain trace elements or micro-nutrients become limiting.

New knowledge and analytical techniques have for example shown the need for the trace element selenium, especially in certain areas of the country where it is lacking naturally in fodder. The nutritional requirement for selenium in the diet is an interesting one. Apart from the very low levels required (of the order of 0.1 mg selenium per kg DM) there is a complex interrelationship with vitamin E, both being involved in preventing a wasting disease of calves called white muscle disease.

Vitamin E in the rumen can be destroyed by oils which are rich in polyunsaturates such as fish oils. The inclusion of cod liver oil in the diet as a good source of vitamins A and D can lead to muscular dystrophy because the vitamin E level in cod liver oil is insufficient to

protect against the effects of the unsaturated fats in the oil, and additional vitamin E is required.

Selenium is toxic to the calf at levels not much above the maximum required so very careful mixing into the feed is necessary and operators also need to take appropriate safety measures when handling selenium compounds.

In addition to selenium there are about seven other trace elements required in the diet at levels measured in parts per million (or grams per tonne). The list includes copper, iron, manganese, cobalt, zinc, iodine and sulphur. Iodine deficiency is well known, affecting humans also, especially in some parts of the country (Derbyshire). Iodine is needed by the thyroid gland to produce a hormone connected with growth and the regulation of the calf's metabolism. Calves can be born suffering from iodine deficiency (goitre) through a lack of iodine in the diet of their dams.

Zinc deficiency is rare and is associated with skin condition. Iron and copper are needed for making the haemoglobin in blood. There are copper-deficient areas in Great Britain where progressive loss of condition occurs unless copper supplementation is given. Calves fed exclusively on whole cow's milk are liable to develop anaemia because of the lack of iron. This condition was also often induced in veal calves by not including iron in the milk replacer, in order to produce pale flesh. However this now is contrary to the code of practice for ruminants.

The function of cobalt is concerned with the synthesis of vitamins B12 by rumen micro-organisms.

Sound recommendations for the requirements of all these minerals are available in the Agricultural Research Council's *Nutrient Requirements for Ruminant Livestock* (1980). The manufacturers of both milk replacers and dry concentrate diets should provide adequate minerals to meet these requirements. Farmers who home-mix can meet requirements by using specially designed mineral pre-mixes to supplement the minerals occurring naturally in feeding-stuffs.

Vitamins

Besides vitamins D and E already briefly referred to, the calf's other main requirement is for vitamin A, which the rearer associates mainly with calf health. The calf is born with practically no reserves of vitamins A, D or E in its liver and is very dependent on the rich supply contained in colostrum. Conscious of the fact that all calves may not receive adequate colostrum and that levels of vitamin A may be low when cows calve at the end of the winter, manufacturers of milk replacers put in high levels of this vitamin which are several times the maximum requirement. Vitamin levels per kg are declared on the bag ticket. Typical good levels would be 40–50,000 international units (i.u.) of vitamin A in a protected form designed to remain potent for at least six to nine months. Levels of vitamin D are about 7,000 i.u. and vitamin E 25 i.u. per kg.

The rearer may be confused by references to pure vitamin A (also called retinol) in terms of 1 microgram (μg) which is equivalent to 3.33 i.u. It is also equivalent to 6 micrograms B carotene which is the precursor of vitamin A found extensively in fresh green crops and well-preserved hay and silage. One decision the calf rearer has to make is whether it is necessary to undertake any further supplementation of vitamin A when purchased calves arrive on farm. A deficiency of vitamin A is associated with diseases like pneumonia, severe scour and eye conditions especially when calves are stressed when moving from their farms of birth. Some rearers argue therefore that it is a positive approach to disease prevention to drench with vitamin A when the calves arrive in order to build up liver reserves quickly. Others rely on the daily intake from the milk replacers and save the cost of the extra supplement. It is entirely a personal decision since no real proof exists either way.

The pre-ruminant calf is not able to synthesise its requirement of the complex of B vitamins, and these are also normally added to milk replacers. Later, when the calf is ruminating, it is capable of supplying its own B vitamins and therefore they are not added to dry concentrate feeds. However cobalt is still needed to synthesise vitamin B12 in the rumen. Occasionally one well-known B vitamin, thiamin, can be destroyed in the rumen or its absorption interfered with and the deficiency would cause symptoms of blindness and uncontrolled staggering in circles. The condition is known as cerebrocortical necrosis (CCN). The calf usually responds dramatically to injections of thiamin if the condition is diagnosed early.

Vitamin C is also often included in milk replacer diets and vitamin K can also be supplied.

Milk Replacer Powders

In their simplest form the milk replacers which calf rearers have used for many years are mainly formulated from by-products of the dairy industry, in conjunction with animal and/or vegetable fats. The two main milk products are dried skim milk powder resulting from butter production and whey powder resulting from cheese production.

Skim milk powder (SMP) is rich in milk protein (casein) with a level of about 35 per cent, and its value is supported in the EEC through an 'intervention price'. Further there is a 'subsidy' on its inclusion in calf milk replacers known as Feed Aid. This is payable only if SMP constitutes a variable amount of the complete milk replacer, when this will be stated on the product bag or label. Much of the liquid skim milk produced in the United Kingdom, Eire and on the Continent has fat added into it to restore the energy lost when the butterfat was removed. Such a product is known as fat-filled milk powder and is commonly used as a means of adding fat to milk replacers. The fat is normally homogenised with liquid skim before spray drying and a free-flowing, non-greasy powder, which mixes easily in either warm or cold water, is produced. It consists of small globules of fat coated with milk powder.

Easy mixing is further enhanced if the drying process produces the powder in an 'instantised' form. This is obtained by retaining an agglomerated form in fluidiser beds after the main drying chamber and coating the product with lecithin. Alternatively fat can be added directly to skim milk powder in the blending process used in the milk replacer manufacture. Although nutritionally this appears to give satis-factory results, it produces a powder which tends to lump in store and mixes less well.

Such powders usually have to be reconstituted using hot water, and because they are not instantised they require much more effort to mix. Whilst this may be a drawback when mixing by hand in the bucket, it is not so important when bulk mixing is carried out in a mixing machine. Even with good mixing in hot water, followed by cooling to blood heat, the fat may not be so well digested because its droplet size is much larger than that of good-quality fat-filled powders.

One cannot be dogmatic about this sort of product and much depends on the quality control and on the expertise of the manufac-turers which is increasing all the time. Normally a milk replacer produced by direct fat addition would have a lower production cost than one based on fat-filled products.

Normal whey powders are low in protein at a level of 12–13 per cent, but they are high in milk sugar (lactose). Because they are a lower-value product and because cheese production varies with the numerous types produced, whey powders have not, in the past, been so well controlled for quality and consistency. However this has changed in recent years and consistent good-quality whey powders are available as well as many new types of powder with higher protein content. Fat-filled wheys are also becoming available although these are normally produced on the Continent rather than in the United Kingdom.

There is no intervention price or 'Feed Aid' on whey powders which, as well as usually being included in 60 per cent SMP milk replacers at a relatively low level, form the basis of milk replacers with little or no SMP usually referred to as zero powders.

If no 'Feed Aid' is paid for SMP inclusion, it becomes an expensive protein source and other proteins to supplement the whey protein are used. The main replacements for milk protein in calf milk replacers are based on products such as suitably processed soya, fish, single-celled synthetic proteins and high-protein wheys, either delactosed and/or demineralised or whey protein concentrate. Blood products are also being produced in France to replace milk protein, and these appear to have a good amino acid profile and perform well, though unattractive in appearance. The advent of BSE has effectively ruled out these blood products.

The processing involved not only includes extremely fine grinding of products like soya, but complex chemical processing to render them more suitable for digestion by the calf.

The protein from such sources does not form a clot in the abomasum in the way which milk-based products do, and digestibility is not normally as good as with the casein in milk or skim milk powder. Nevertheless the calf's digestive system seems to be able to adapt itself and many thousands of calves are being reared very satisfactorily on replacers based on whey and substitute proteins, and a cost saving of £400 per tonne (or £5–£10 per calf), which is a significant amount, can be made.

It is generally agreed that these products do not work as well under stress conditions with purchased calves, or where management is less satisfactory. Also it has been shown that the ability of the calf to digest non-milk products improves rapidly with age. Therefore a home-bred calf given a good start on whole milk can utilise such proteins very well at a later stage. Not only will protein digestibility normally be slightly lower, but total digestible energy supply will be reduced which will encourage a greater dry feed intake by the calf in the five weeks up to weaning. Total intake to five weeks could be expected to be increased by 3–4 kg per calf. This higher intake helps the weaning process and enables the calf to keep growing without check at this time. Again the expertise of the manufacturers is the key to the success of any particular product. There have been tremendous improvements in the last five years in the techniques of raw material processing availability and evaluation.

Acidified Milk Replacers

The acidified products were introduced into this country in 1978 and were designed for the cold ad-lib system. The acids used, it should be stressed, are mild organic acids and not strong mineral acids.

The keeping time in the containers is a function of acidity level in the powder, temperature of the liquid, and the quantity which is still in the container when it is 'topped up' with a fresh supply. The biggest variable here is the ambient temperature and there is an obvious need to keep the containers out of direct sunlight. Similarly, in winter some protection from freezing temperatures is essential because, although the product will keep well, intake of milk replacer below 10 °C will be reduced. Frozen pipe lines will of course reduce consumption to zero.

Acidified milk replacers can be based on formulations containing skim milk powder or on 'zero' diets. The former can only be acidified to a moderate level if casein precipitation is to be avoided on mixing. Their pH will be about 5.7 at which level there is no perceptible effect on palatability to the calf. They would usually have an oil content of 17–18 per cent and protein of 23–24 per cent and be high in energy.

Normal keeping time will be a minimum of three days, which means that quantities used should be adjusted so that bulk containers become empty every third day and can be cleaned out. If cleaning is delayed there is a danger that normal souring would accelerate, especially in hot weather, and the casein would coagulate into a curd at about pH 5.0.

With milk replacers based on non-milk protein the problem of casein precipitation does not arise. Consequently a higher level of acidity can be used if desired, and this gives greater flexibility over cleaning-out times. However, this replacer may be less acceptable to the calf, especially during the initial training on to the system. Not all zero diets have a high acid level however.

Milk Replacer for Machine Feeding

One attribute required in milk replacers which are suitable for most types of machine feeder is that they must flow satisfactorily. Apart from this such products are of similar composition to those used for bucket feeding. Because the hoppers and feed mechanisms differ from one machine to the next, manufacturers of milk replacers face an impossible task in providing a single powder which will suit all types of machine. There is a tendency therefore for there to be a close link between machine and milk replacer manufacturer.

Instantised milk replacers (based on instant fat-filled skim-milk powders), which are the main products used in the United Kingdom, do not flow well and are normally not suitable for machines, even where a vibrator on the hopper is available. The flow properties of such milk replacers can be improved by the use of a suitable flow agent, usually some form of silicate, and thereby can be made suitable for machines.

Milk replacers imported from the Continent usually consist of non-instantised ingredients, and some of these flow satisfactorily in machines without the addition of flow agents. However those with fat added directly (which retain their shape when squeezed by hand into a ball) are generally unsuitable for machines.

EARLY WEANING DRY COMPOUNDS

No attempt will be made under this heading to provide formulations, or to list suitable straight raw materials and micro ingredients for early-weaning diets. This is because survey work shows that, on average over the country, 87 per cent of all calves are given a compounded feed initially in the rearing programme. This is understandable because such compounds are highly palatable and research based. There is also the difficulty of buying ingredients economically, of storing a relatively large number on farm and keeping them fresh. There is also the problem of mixing and pelleting such diets, especially to the preferred small size of 3–5 mm. Additionally it is essential with coarse mixes to keep dust to low levels. Such diets usually have molasses sprayed on partly to keep dust down but also to increase palatability.

Expert nutritional advice is available to anyone who wishes to home-mix early-weaning diets, but even specialist rearers who are large users tend to purchase compounded diets and to save costs by buying in bulk and blowing into hoppers. Of the compounds purchased there is a ratio of about 3:2:2 for an 18 per cent protein pellet, a 16 per cent protein pellet and a 16–18 per cent protein coarse mix feed. Dairy farmers and specialist rearers tend to prefer the 18 per cent protein pellet to the 16 per cent. Rearers of beef calves use the coarse mix the least, but coarse mix users keep calves on the diet longer, feeding up to eleven weeks of age compared with the pellet feeders up to ten weeks.

Once past the critical stage of rearing there is a greater use of home-grown cereals and sugar beet pulp balanced with protein concentrate pellets to form a 14 per cent crude protein diet. This is most marked amongst specialist rearers and beef men, whilst about half dairy calf replacements go through a post twelve-week period on a 14 per cent protein calf-rearing compound.

Pellets Versus Coarse Mix

The debate whether pellets or coarse mixes are better usually generates quite a lot of heat amongst calf rearers. The controversy has certainly been going on for some time and it is not likely to be resolved easily. Views are very entrenched and each has its champions.

Ingredients for coarse mixes have to be carefully chosen and each has to be palatable in its own right to prevent the calf from selecting individual raw materials. Even so in every batch of calves being individually reared at Barhill there are three or four calves in a hundred which will reject one particular ingredient. With some it is linseed lozenges, with others it may be locust bean. The majority of calves take the complete mix as intended, but the most common fault in a coarse mix is too much dust or meal particles. This is caused by over-vigorous mixing and the resulting meal is generally rejected by the calf. Ingredients like maize, barley and peas are usually steam flaked or micronised, and these together with high-protein cakes give the mix an open and varied appearance which is usually attractive. Flaked cereals are more expensive than the same materials in ground form, and coarse-mix diets usually cost considerably more than pellets. Typically 30–40 per cent of a coarse mix will consist of pellets which contain ground high-protein feeds, principally soya, plus all the minerals, vitamins and any growth promoter included. Finally the whole of a coarse mix is often sprayed with molasses at about 7–10 per cent.

When it comes to intake and performance there are no startling differences between coarse mixes and nutritionally equivalent pelleted diets. To make a true comparison the same formula should be fed in coarse form and in pelleted form. When this has been done at Barhill there has normally been a small performance benefit in favour of pelleting. In some comparisons coarse calf feed has been consumed at a slightly higher level than pellets in the pre-weaning stage. This has not always been accompanied by an equivalent increase in liveweight gain however. The loss in efficiency has been blamed on the rapid fermentation of the flaked cereals in the rumen which in turn has affected the absorptive efficiency of the rumen wall.

With the cost up to £50 per tonne higher for a coarse mix compared to a nutritionally equivalent pelleted product, the rearer has to decide for himself, given his own circumstances, whether he feels this extra cost is justified. The additional cost will be of the order of £7.50 per calf reared to 100 kg liveweight.

Pelleted Diets

As previously stated the current preference by calf rearers is for an early-weaning diet in small pellet form. This is helpful to manufacturers since a reasonably hard product, free from dust, can be made without adversely affecting intake. Larger pencils, if made too hard, will certainly depress intake in the early weeks of a calf's life.

Careful selection of palatable ingredients, and a restricted number of them allowed in the specification of the first diet for young ruminants is necessary if satisfactory intakes are to be achieved. It is quite easy to demonstrate that even a 5 per cent inclusion of a less palatable ingredient will depress intake quite severely. One of the main advantages of a pellet is that each is a complete formulation and therefore no selection of ingredients is possible. This is important because it allows the new developments in the precise formulation of ruminant diets to be implemented.

Chapter 4

CALF HOUSING

Types of Calf Housing

THERE ARE basically three types of calf housing to satisfy three production systems: (1) for dairy and beef calves reared as replacements or for beef production, (2) for veal production and (3) for suckled calves. Most calves reared come under the first category and this chapter concentrates on their requirements. The spatial requirements for each category are different and it is essential for advisers to be clear regarding the form of housing required.

Advice on Calf Housing

There are several sources of advice available to farmers today, and when seeking calf housing advice the following simple questions will establish the adviser's experience:

(a) Has the adviser ever reared calves?
(b) Has he/she designed other calf houses and if so did they work satisfactorily?
(c) Is the person giving buildings advice a veterinary surgeon?

If the answers to (a) or (b) are no be very careful indeed. If the answer to (c) is yes again be wary. Would you ask a buildings expert for veterinary advice? Clearly no.

Another problem appears to occur when pig or poultry specialists are let loose on a calf house. Regularly they impose the precise environmental requirements of pigs or poultry without establishing clearly that a calf is different both in shape and composition from either a hen or a pig!

Basic Objectives

What then are the housing options for calf rearers and how does the rearer evaluate a design?

The main objective in the planning and design of calf housing is to provide an environment which will minimise the requirements for veterinary treatment, minimise calf mortality and encourage the production of healthy calves.

Calf housing should provide a suitable environment for the calf and the stockman. It should also provide ample space for husbandry and management tasks and should be planned in detail to achieve these objectives with the lowest possible capital and running cost and with a low labour requirement. One building up to twelve weeks of age minimises problems and checks in growth rate which occur when two buildings are used.

In the past farm building designers and manufacturers of package-deal buildings provided a similar structure for poultry, pigs and calves without ensuring that it was suitable for calves. Most farm buildings are designed by surveyors or architects. They often lack agricultural experience and as a result tend to adopt an anthropomorphic approach. This has resulted in fan-ventilated insulated structures for calves with provision for supplementary heating.

A fan-ventilated, insulated structure is *not* required for calves. Experimental studies carried out throughout the United Kingdom including comparative housing studies near Aberdeen, near Ayr and at Ministry of Agriculture Experimental Husbandry farms have established the feasibility of well-designed naturally ventilated uninsulated housing for calves from birth to twelve weeks of age and beyond. This form of housing is suitable for calf rearing throughout the year. The experimental work has been carried out over the last twenty years and today the majority of calf houses are naturally ventilated. Package-deal building manufacturers have followed this trend and produce naturally ventilated calf houses.

It is interesting to note that calving boxes are usually naturally ventilated and uninsulated with no supplementary heating. If these conditions have satisfied the needs of the calf at birth how strange it seems to have provided warmer conditions for a healthy calf beyond this stage.

The main advantages of naturally ventilated calf housing are:

1. They provide the required environmental conditions for calf rearing. The calf, unlike (for instance) a young pig, can tolerate a wide range of conditions without the cost penalty of a poor economic response.
2. A reduction in capital cost of 20–30 per cent when fully costed against insulated fan-ventilated structures.
3. Correctly designed—a reduced management requirement because

the ventilation system requires no adjustment throughout the year. The stockman can concentrate solely on calf rearing. The minimum ventilation rate depends on the stack effect and the maximum on the wind effect:

Stack Effect Ventilation Rate—The ventilation rate on still days due to warm air rising by convection. Air leaves by outlets in the ridge and enters via openings at eaves level.

Wind Effect Ventilation Rate—The ventilation rate due to wind pressures on the outside of a building.

Always design to achieve a satisfactory stack effect ventilation rate which ensures an adequate ventilation rate on still days, then check the wind effect rate and the internal airflow patterns.

4. Low running costs (no heaters or fans).

Whole house heating is not used for calves simply because it is not required, i.e. no benefit would be obtained. However this does not mean that local heating is never used. For single sick or weak calves electric infra-red strip or panel heaters should be used over individual pens or in a separate pen where calves are group housed. Infra-red lamps should be avoided due to the fire risk should the glass be broken over straw bedding. Always include electric sockets so that heaters or pressure washers can be used.

Calf housing must provide a high natural ventilation rate and large air volume for animals. This is ideal during most of the year, but how can heat be effectively provided under these conditions for a sick calf or under damp conditions during the autumn and winter months? The ideal heater would be mounted high above the calves yet heat the calf and not the air within the building.

A new electric heater, the quartz linear lamp (QLL), is making a major contribution in calf housing for single and group-reared calves. It has a fast directional response with an efficient long-throw performance. The beam penetrates damp air with minimal atmospheric absorption to heat calves which enter the beam rather than the air in the calf house.

A rearer in Yorkshire first installed one 1½ kW QLL heater over each pen of thirty calves in November 1987 and found that calves slightly off-colour or sick immediately moved into and stayed in the heated area while healthy calves kept out of it. His results have been very impressive: 'My veterinary bills have dropped by 75 per cent and the installation and running costs were recovered in the first three months! The QLL heaters are a "free vet" and we are incorporating them over each grouped pen on our new farm. At last there is a heater which can be

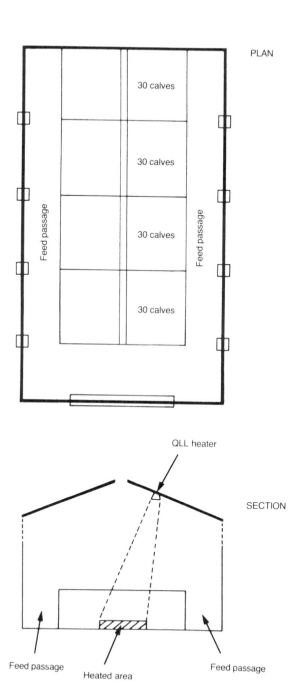

PLAN

30 calves

30 calves

30 calves

30 calves

Feed passage

Feed passage

QLL heater

SECTION

Feed passage

Heated area

Feed passage

used to heat calves while maintaining specific air volumes, ventilation rates and draught-free conditions in calf houses.' Since 1987 the use of QLL heaters has been spreading rapidly.

The work and experience with naturally ventilated calf housing has shown that adequate ventilation and air volume, not temperature, are the main requirements of calf-rearing houses. This is as long as the air movement rate close to the calves is checked, particularly during the winter, to ensure that air speeds at calf level are below 0.25 m/s (50 ft/min). In a calf house in winter it should not be possible to feel air movement at calf level on the back of the hand. The major emphasis on temperature in the past was probably because designers felt that heat loss depended mainly on air temperature. Awareness that heat loss depends on a combination of air temperature and air movement rate coupled with the knowledge of the wide range of conditions suitable for calves has assisted in the acceptance of naturally ventilated buildings. Naturally ventilated housing is now widely used for calves reared as dairy herd replacements, for beef and for veal production.

When converting a building for calves it may prove impossible to achieve the required environmental conditions with natural ventilation. In this case fan ventilation can be installed but *manual* fan control and not thermostatic control should be used. Whenever thermostatic control has been installed it is common to find no ventilation during the winter months due to erroneous emphasis on temperature. This defeats the requirement for good continuous ventilation. As temperature is not important it follows that insulation is not required in the roof or walls.

At the start of any buildings design exercise the design requirements or brief should be clearly established and the design developed to satisfy this brief. Problems on farms can often be traced back to a misunderstanding between the farmer and the designer due to lack of preparation of the initial brief. If there has been a change in rearing policy or stocking rate this can be established clearly by comparing current practice with the initial requirements. Increasing the stocking rate over time can lead to problems because it reduces the cubic air capacity (or air volume) and ventilation rate per calf, increasing the dangers of cross infection. It also increases the moisture load in the building.

Calf accommodation should be designed to provide a minimum disease break of three weeks between batches. Problems have occurred when the planned disease break of three weeks has been reduced to below even one week. This can result in problems after years of successful rearing with a particular system.

What are the fundamental requirements we are trying to satisfy in calf housing? The designer is basically aiming to provide:

- a dry bed
- a well-ventilated environment
- a specific minimum cubic air capacity (or air volume) per calf
- a draught-free environment at calf level.

It is easy to list these requirements. What is the reason for each one and how do they relate to a design exercise?

A Dry Bed

A dry bed is important to reduce heat loss to the floor and minimise straw usage. A dry bed is provided by giving adequate floor drainage beneath straw bedding, and a minimum floor slope of one in twenty is required. Straw bedding will not move down this slope. If long slopes are required they can often be split into two slopes to reduce the drainage distance and concrete required. The effect of flooring type on the calculated lower critical temperature for an 80 kg veal calf in calm air on full feed illustrates this point in table 4.1. The lower critical temperature may be defined as the air temperature below which the calf will start to direct energy from growth into keeping warm. It is important to remember that the lower the critical temperature the better the calf is able to resist cold stress, and from table 4.1 it will be seen that dry straw is worth about 5 °C compared with wet straw bedding.

Table 4.1   Calculated lower critical temperatures for a calf* in calm air on full feed

Relationship to floor	Critical temperature (°C)
Calf standing	−3
Calf lying on dry concrete	−6
Calf lying on dry straw	−8
Calf lying on wet straw	−3
Calf lying on wooden slats	−3

*Veal calf (80 kg, 150 w/m²)
WEBSTER, A. J. F., GORDON, J. G., McGREGOR R., The cold tolerance of beef and dairy type calves in the first weeks of life, *Animal Production 26* (1).

It is often necessary to have close site supervision during construction to ensure adequate floor slopes and good internal drainage layout. Water bowls, automatic feeders or feed buckets *must* be placed at the lower end of sloped floors to ensure that bedding remains dry. Moisture removal from a calf house is usually by a combination of

Plate 13   Bad drainage layout. Would you like to work in this calf house?

drainage and ventilation. Poor drainage results in an extra load on the ventilation system. Passages in calf houses must be kept dry. A tendency in the past to wash these down daily simply added moisture to the building, placing an extra load on the ventilation system.

Good Ventilation

In naturally ventilated calf houses whether of pitched roof or mono-pitch section, the minimum ventilation rate is calculated on the basis of the stack effect and the maximum on the wind effect. Ventilation systems *must* be designed to produce specific rates. Design data summarised at the end of this chapter will ensure an adequate minimum rate per calf. Ventilation should be designed to remove moisture vapour produced, disease organisms, dust and foul air, and replace them with fresh air. Ventilation should also remove by-products such as ammonia, hydrogen sulphide, carbon dioxide and methane.

Cubic Air Capacity

The cubic air capacity per calf is important in all calf housing designs because:

- it dilutes the intensity of disease organisms in a building, thus reducing the dangers of cross infection;
- height and space allows air to be introduced to a calf house well above the level of calves thus minimising the risk of draughts at calf level during the winter months.

When designing a calf house the layout planning leads on to the roof and wall heights required. At this stage cubic air capacity per animal should be checked to determine roof and wall heights, which should be checked again to ensure that heights are suitable for tractor access and husbandry tasks. Approximately twenty years ago it was noticed that respiratory problems were not uncommon in buildings with cubic air capacities of about four cubic metres per calf. They proved very diffificult to improve by increasing ventilation rate but were generally improved if stocking density was reduced. This increased the cubic air capacity per calf and a minimum of six cubic metres per calf proved acceptable. In the last few years Professor John Webster and his team at the Bristol Veterinary School have looked in detail at air hygiene in calf houses and have provided the scientific background to emphasise the importance of cubic air capacity. Both ventilation rate and cubic air capacity *must* be defined in a design.

When converting buildings with a lower roof, such as a poultry house, to calf accommodation the layout plan may indicate that thirty calves can be penned in the given floor space. However this often results in inadequate cubic air capacity per animal. In these cases check both figures and accept a lower stocking rate if cubic air capacity is limiting.

A Draught-free Environment

Model Study of Airflow in Two Calf Houses
The internal air flow pattern in a calf house is very important. Model studies were carried out to determine the general airflow patterns inside calf houses and examine the effect of pen covers together with solid and open pen fronts on the air flow patterns in two naturally ventilated calf houses. One calf house was a model of an experimental climatic calf house with two rows of pens and a central passage (figure 4.1). Pen covers were first used over the rear of the pens in 1968 and continued in use throughout the work. The other had four rows with

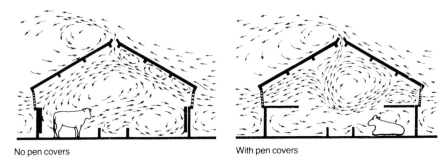

No pen covers   With pen covers

Figure 4.1   Effect of pen covers on airflow in a two-row house

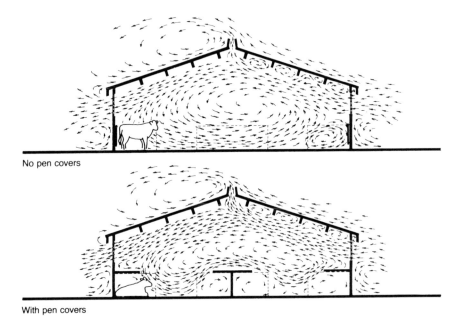

No pen covers

With pen covers

Figure 4.2   Effect of pen covers on airflow in a four-row house

two passages and was a typical four-row naturally ventilated calf house (figure 4.2). In both buildings air inlets were below eaves level and air outlets were open ridges.

The flow patterns studied were based on flow in the plane of a cross-section from right to left. This flow represented the wind effect component of natural ventilation. The stack effect was ignored together with the effect of heat output from the calves. For both buildings the various combinations of pen cover and pen type were examined. From the range of combinations used a selection has been made to illustrate the main points arising.

Pen Covers

Model tests were carried out for both the two-row and four-row calf house designs with and without pen covers over the rear of the calf pens. Omitting the pen cover results in a downcurrent on the wall opposite to the main air inlet and part of the primary air flow occurs at calf level.

The presence of pen covers raises the lower limit of the primary air flow, thereby effectively reducing the air speed at calf level. Secondary, slower flow occurs beneath the cover at calf level. In a full-scale calf

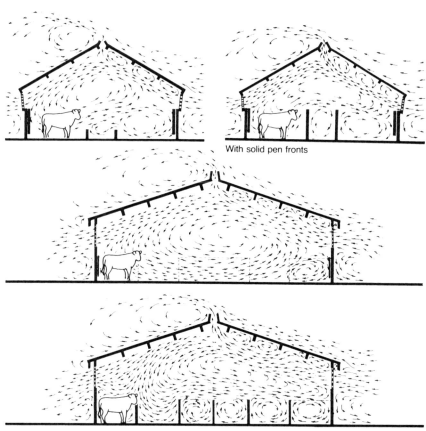

With solid pen fronts

Figure 4.3   Effect of pen fronts on airflow

house this secondary flow together with convection above the calves ensures adequate ventilation at calf level.

The pen cover has a similar effect when the air inlets are representing space boarding or when they represent openings between the pen cover and the eaves. The latter situation might be used under summer conditions if the ventilation panels can hinge downwards.

Solid Versus Open Pen Fronts

Solid pen fronts also reduce air speeds at calf levels but do not prevent downcurrents along the walls (figure 4.3).

Solid Fronts with Pen Covers

In model tests where both pen covers and solid fronts are used the pen cover tends to have the dominant effect as far as raising the primary flow and thereby reducing air speed at calf level is concerned.

The main effect of the cover is to prevent downcurrents of air along the sides of a calf house. It is primarily a flow deflector and reduces air speeds at calf level.

There is no need to insulate pen covers. Erroneous statements regarding covers have implied heat conservation but the main aim of a cover is airflow direction. There is also no need to cover any more than the rear 800 mm of a pen.

Temperature patterns recorded in a climatic calf house show that temperature lift beneath a cover is small in a well-ventilated calf house (table 4.2).

Table 4.2   Mean temperatures recorded in a naturally ventilated calf house above and below pen covers

Batch	Week	Temperature above pen cover (°C)	Temperature under pen cover (°C)
A	6	9.4	11.4
A	7	5.7	8.1
B	6	17.7	19.3
B	7	14.4	16.2
C	6	8.9	10.3
C	7	9.7	11.4

In this naturally ventilated calf house, air speeds at calf level did not exceed 0.25 m/sec (50 ft/min) below the pen covers. Above the covers speeds up to 1.25 m/sec (200 ft/min) were recorded in winter. But this was well above the calves.

From the model studies outlined, both pen covers and solid pen fronts have an effect on the flow patterns in calf houses and therefore an effect on the air speed at calf level.

The pen cover has a more marked effect than solid pen fronts. It is also interesting to note the low air speeds recorded at calf level during the winter months in a naturally ventilated calf house with pen covers and with railed rather than solid pen fronts.

A completely solid pen front would have a serious disadvantage in practice as feed buckets would have to be inside the pen where they can easily be fouled. However partly solid pen fronts which allow calves access to buckets outside the pen are possible.

If solid pen fronts and solid pen divisions were used calves would not be able to see each other which is undesirable. In certain cases it might be useful to incorporate an occasional solid pen division, e.g. to prevent cold draughts in winter along the length of a calf house. The freedom of choice regarding solid and open pen fronts and divisions should be left to the discretion of the designer concerned. Simple model studies could be used to indicate the effect of different internal designs.

Model Study of Airflow in a Wide-span Monopitch Calf House
At the planning stage of a monopitch complex for calf housing it was apparent that decisions relating to siting and details could not be satisfactorily made without results from model studies. With a two-row monopitch design it was important to examine the effects of:

1. facing the units towards each other;
2. facing both units in the same direction;
3. adjusting the gap between the units;
4. having solid gates and downhung sheets on the front, and space-boarding on the low side (back) of the monopitch.

The results are shown in figure 4.4.

The models were positioned in three different ways relative to the wind: the fronts together, the fronts facing the prevailing wind and the backs of the building facing the wind. The flow patterns studied were again based on flow in a cross-sectional plane from right to left. This flow represented the wind effect component of natural ventilation. The stack effect (the effect of ventilation due to heat output from the calves) was ignored.

From the flow patterns developed from each of the arrangements it was clear that spaceboarding to a depth of 0.6 m on the end wall gave adequate air movement in each of the houses. With the back wall solid the building on the windward side showed no movement within.

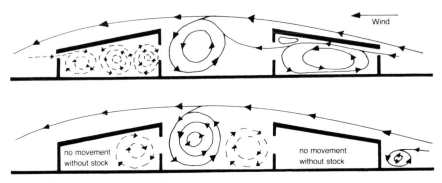

A   The effect of an opening in the back wall

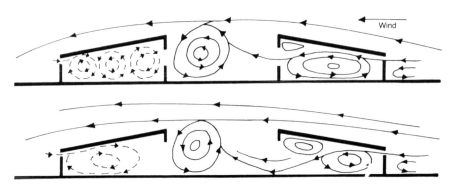

B   Solid and open front gates

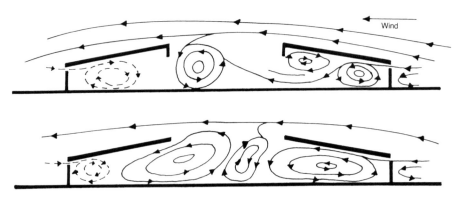

C   The effect of a top skirt

Figure 4.4   Model study of airflow in a wide monopitch house with full height divisions between sections

However, in this situation the heat from the stock would produce some air movement by convection (A).

The model buildings with spaceboarding were then arranged in three different positions to examine the need for a solid or open front gate (B). No significant change in the general flow patterns was evident and the downward eddy between the buildings did not enter the building on the lee even when the gate was removed.

The need for a top skirt was then examined (C). A downward roll from the top resulted in fast movement within the second house. In the building with its back facing the wind, circulation was adequate. The buildings with the top skirt had flow of a nature not detrimental to any stock. The skirt also helped to stop driving rain entering the building on the lee.

Conclusions
1. In a large-span calf house 9 m wide a spaceboarded opening on the back wall is preferred for adequate air circulation. However in a short-span calf house (4.5 m) this is not needed.
2. There is no indication that a solid or open front gate affects the general flow circulation within the buildings. Flow patterns studied were based on flow in a cross-sectional plan from right to left, not onto the gable. There would be no point in making them solid except to prevent the entry of driven rain.
3. A top skirt on the front would prevent considerable down draughts in the building on the lee. It would also help to prevent driving rain from entering the building.
4. The evidence suggests that the buildings could be positioned in two ways, either both fronts facing together or both backs facing the prevailing wind.

Relative Humidity (RH)

The RH in a naturally ventilated calf house is slightly lower than the external RH due to the slight temperature lift inside the building over the outside temperature and the high ventilation rate. It follows the outside RH pattern closely so is usually low during daylight hours as air temperatures rise during the day. The air temperature under covers is slightly higher than the air temperature in the middle of a calf house so RH is even lower beneath pen covers.

There is currently active discussion on RH relative to the number of disease organisms in the environment of a calf house and whether or not the number can be positively related to specific respiratory problems. Whether RH or absolute humidity should be the main factor is not yet clear. Controlled environment calf houses are not required for calves but in the past resulted in a continuous high RH (80 per cent or more) which, combined with high air temperatures encouraged rapid multiplication of disease pathogens. Insulating a naturally ventilated calf house has little effect on air temperature due to the high ventilation rate used and low heat output from calves. The RH is slightly depressed following insulation but continues to follow the external conditions. There is no evidence to show that this marginal change in average RH has any effect on the calf's health or performance. If any improvement results it would have to be accurately quantified and balanced against an increase in capital costs of approximately 8–10 per cent for insulated over uninsulated calf accommodation.

Evidence that disease organisms breed in straw bedding may mean that improving the aerial environment may be the least effective way of reducing the number of disease pathogens in a calf house. Sterilising the bedding may be a better way and new techniques for doing this, such as the Electricity Council's Radio Frequency sterilising system, need to be examined in detail on a cost/benefit basis.

The best advice for farmers at the moment is to avoid a continuous high RH by providing an uninsulated, well-ventilated calf house with a high cubic air capacity per calf and draught-free conditions at calf level during the winter months.

Condensation, Rain and Snow

For the same reasons, condensation should not be a serious problem when the roof is uninsulated. The temperature lift from inside to outside is only about 2 °C and the high ventilation rate removes most of the moisture produced.

The question is sometimes asked whether open ridges allow a lot of rain in. When correctly calculated the ridge gap is narrow and should not admit much moisture, and at the same time the amount of moisture leaving by the ridge is considerable. If site conditions result in snow entering, a raised ridge cap can be added but it is not necessary to put a ridge cap on from the outset.

HOW TO CONVERT EXISTING BUILDINGS INTO CALF ACCOMMODATION

Unsuitable Buildings

Most new calf accommodation is based on converting existing farm buildings and many are successful for calf rearing. I am basically looking for buildings which satisfy the calf's needs for space; cubic air capacity per animal; ventilation; drainage, and access to reach and service group or individual pens. From experience with conversion work it is also important to be sure a building is suitable for conversion. Which buildings should be avoided? The following list indicates which buildings have been generally *unsuitable* for conversion and if used often result in problems.

1. Pitched Roof Wide-span Buildings (for example 12 m (40 ft) wide and with low pitches).
In this type of building it is difficult to provide adequate draught-free ventilation for calves either naturally or mechanically and it is tempting to have too many calves (sixty or more) in one air space. If individual pens are used there is a danger of a high stocking rate leading to an inadequate cubic air capacity per calf.

2. Buildings Jammed in between Other Buildings
Again these can be difficult to ventilate. It can also be difficult to provide adequate drainage particularly if surface drains have to run through adjacent buildings. If the surrounding buildings are stocked with older cattle there can be problems with air moving from older cattle to young calves. Access for all activities including mucking out can be very difficult. I usually advise farmers to avoid this situation for calf housing. It is often better to build a new free-standing unit well away from existing buildings and if detailed costings are examined it might even be cheaper.

3. Lean-to Problems
Free-standing monopitch buildings make excellent calf houses but whenever a lean-to is used (see figure 4.5), ventilation is difficult to provide because the taller building blocks the high side of the lean-to.

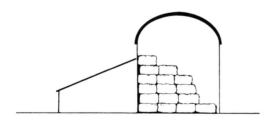

Figure 4.5   It is difficult to achieve ventilation in a lean-to

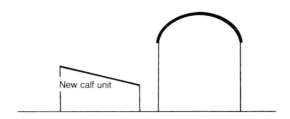

Figure 4.6   Build a free-standing monopitch

There is also often a limit to the height of the building so cubic air capacity becomes a problem and there may possibly be too little access for mucking out by tractor. At the planning stage do not add a lean-to on to an existing building to create a calf house. It is far better to create a separate building away from the barn (figure 4.6).

If a lean-to has to be used it might be possible to provide adequate ventilation for group-reared calves (but not for individually penned calves) by using air inlets on the low side with a slotted roof. Do not use a slotted roof over individually penned calves in case any moisture dripping through the slots falls directly onto the calves.

4. Poultry and Pig Buildings
Wide-span, low-roofed poultry or pig buildings can be difficult to ventilate and again may provide an inadequate cubic air capacity per calf. If the existing buildings have flat concrete floors for poultry or pigs (e.g. under deep litter in poultry sheds or straw bedding for pigs) then the floor will have to be relaid to provide adequate drainage. This can be expensive and difficult to achieve. Air speeds at calf level need to be checked in these buildings as the existing ventilation systems will almost certainly be inadequate for the calves. Never use poultry or pig buildings with a temperature-controlled ventilation system as this could result in no ventilation during certain times of year.

Good Potential Buildings

The buildings mentioned so far pose problems when considering conversion to calf accommodation and should be avoided. From the difficulties they present it is important to realise that three types of building can make excellent calf accommodation.

1. Free-standing Pitched-roof Buildings
These should be of 6–9 m (20–30 ft) span with eaves, ceiling and/or roof heights adequate for tractor access. In this group if an existing loft

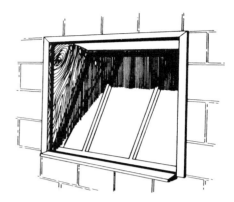

A glazed hopper window like this should be
replaced with space boarding.

ceiling has to be retained successful conversions can be achieved using
manually controlled fan ventilation. This is often simply cross ventilation with internal baffling to avoid draughts at calf level during the
winter months.

Wall space must be adequate to allow air inlets to be created. I
prefer space boarded panels and these can be designed with 12 mm
(½ in) gaps and board widths down to 25 mm (1 in) to allow the
required inlet area if it is difficult to provide a large area of inlet in an
existing opening. Never use hopper-type air inlets for calves. They
allow air in at high speed under windy conditions which can create
draughts at calf level and in practice tend to end up nailed shut!

2. Dutch Barns
It should be possible to provide individual or group pens and add
protection on the sides to prevent direct rain or snow entry but not to
limit ventilation.

3. Free-standing Monopitch Accommodation

New Calf Accommodation—Monopitch Designs

For new calf accommodation a free-standing monopitch design should
always be given first priority. It is a calf building designer's delight. No
longer do questions such as how many calves should there be in one air
space arise as the sections of a monopitch design limit the maximum
number to twenty-four due to the space available. Sections can be
designed to house calves of similar age and the number per air space
usually ranges from ten to twenty-four. An adequate cubic air capacity

per calf and ventilation rate are easily achieved (the ventilation requirements for monopitch designs can now be calculated) and tractor access for mucking out is simple thanks to access on the high side. Capital cost per calf place is lower than for any other form of permanent calf accommodation (for instance 8–10 per cent less than a pitched-roof design). Drainage is easily provided for individual pen or group pen layouts and passages can be included at the high or low side to provide the stockman or -woman with adequate protection from the weather.

The divisions between sections should always be full height to prevent through draughts in winter and provide good separation between calves in adjacent sections. A good level of natural daylight is achieved from the open front and in deep monopitches of 6 m (20 ft) or more the space boarding at the low eaves height. However it is still important to provide natural light by way of the roof sheeting, and 10 per cent translucent sheeting should be included. If the orientation is into the south/south-west, winter sun penetration and summer shade will be achieved.

A monopitch design is basically tamper-proof, i.e. there are no adjustable openings and with careful attention to detail the same design can be suitable for both winter and summer calf rearing. Never provide sliding space boarding or hopper windows in any type of calf accommodation. Adjustable openings are nearly always left shut leading to inadequate ventilation. In the early days with monopitch designs

Plate 14   Good natural and
artificial lighting

(fifteen to twenty years ago) tractors did not have cabs, and roofs were correspondingly lower. As tractors have increased in size and cabs have become common there has been a move to increase the roof height for new designs to provide tractor access. This has led to a positive increase in cubic air capacity per calf which might not otherwise have been achieved.

Accurate data are now available to ensure adequate ventilation rates in monopitch designs and the size of openings at the low and high sides can be calculated. These are the minimum requirements and unfortunately some designs have appeared with openings which have practical limitations. In several instances the front gate has been so high that you could not see over it and the downhung sheeting at the front so low it has been removed by tractor cabs hitting it!

When tractors were first fitted with cabs several cabs went through the roof sheeting on monopitch designs as tractor drivers concentrated on mucking out with a foreloader. Always check the heights on a design relative to the equipment to be used and check that a foreloader reaches the rear wall before the cab hits the roof sheets, lights or purlins!

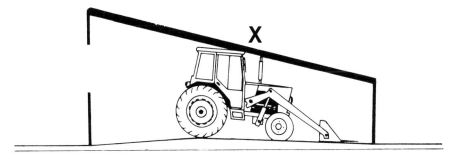

Airflow Patterns in Wide-span Monopitch Calf Houses

To ensure a draught-free environment in monopitch calf houses model studies have been carried out to examine the different design options. The results of this work have shown that the following points are important:

1. A downhung top skirt prevents downdraughts occurring in the building and helps to provide draught-free conditions in the winter. It also reduces the amount of driven rain entering the front and reduces the danger of snow penetrating. (See figure 4.4C.)
2. Solid gates at the front have little effect on the overall pattern but are preferred to prevent driven rain entering at the front.

Plate 15 Monopitch calf house. Height adequate during construction for easy machinery access

Plate 16 Fan-ventilated calf house conversion

Plate 17 Good monopitch design using home-grown timber

Plate 18 Monopitch calf house

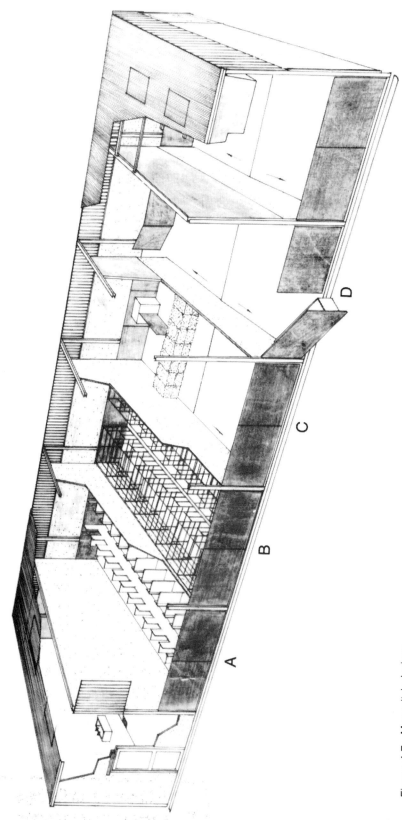

Figure 4.7 Monopitch designs

3. A space-boarded section on the back wall (to form a ventilation opening) is preferred on a deep monopitch to obtain a good ventilation rate through the section.

Monopitch Examples

Each design has to be produced for specific requirements so 'standard' designs are not given for calf houses. However in figure 4.7 monopitch designs A, B, C, D illustrate a number of variations on the basic design for specific requirements.

Monopitch Designs A and B

These sections show individual accommodation as individual pens B or the tethered feed fence A. A would be used for the first three weeks before grouping calves as one batch in section D. B could be used up to weaning at six to seven weeks of age. The calves could then be group reared to twelve weeks in this section by removing the individual pens or they could be moved to a group-reared section such as D.

Both A and B would have an access passage at the front.

A door between sections provides access along a row of monopitch units. The front passage ensures that individual pens are not exposed to driven rain at the front. A solid pen division next to the passage reduces the chance of draughts through the pens.

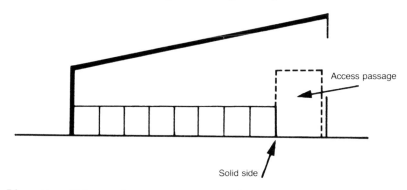
Access passage

Solid side

If section B is used as a group-penned section, post-weaning troughs can be provided next to the front passage or the calves can be allowed to use this space and troughs can be provided inside the solid front gates.

Section C shows a group-penned monopitch section with an automatic calf feeder set into the rear of the pen. In this case the service passage is on the low side of the monopitch.

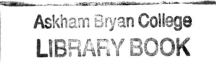

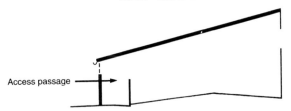

Modern automatic calf feeders can be set in the division between two sections to feed two groups of calves, e.g. two by twenty.

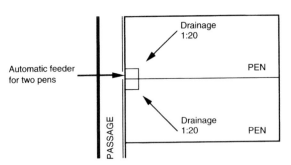

Housing Systems
In all calf housing designs one building (or the same environment, e.g. two sections of a monopitch) should be used from arrival at say one or

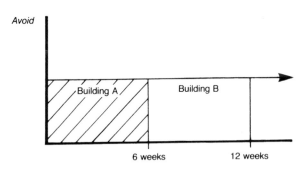

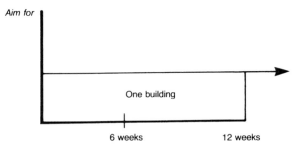

two weeks of age up to twelve weeks of age. This has been shown to reduce health problems and growth checks which were associated with older schemes using two or even three separate buildings (with different environments) up to twelve weeks of age.

To provide the space and volume of air per calf needed one building may seem generous in the early stages. However it is easier to manage, healthier for the calves and cheaper than any two- or three-building system.

PITCHED ROOF DESIGNS

Simple free-standing pitched roof designs should have the following features:

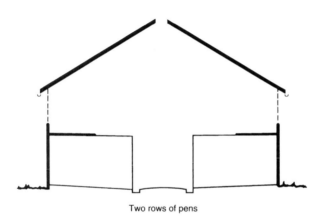

Two rows of pens

1. Adequate floor space per calf.
2. Enough height to allow end access for mechanical muck removal.
3. Floor slopes planned for good drainage of both the pens and passageways.
4. Adequate cubic air capacity/calf.
5. The required air inlet and outlet ventilation areas to ensure good natural ventilation.
6. Uninsulated walls and roof.
7. Air inlets and outlets protected from bird entry: 12 mm (½ in) max air inlet gaps in space boarding and open ridges protected with galvanised 25 mm (1 in) chicken wire.
8. Good natural lighting, e.g. 10 per cent of the roof area as translucent sheeting, and adequate artificial lighting.

Plate 19   Wide-span naturally ventilated house for group-penned calves on automatic feeders

Plate 20   A new 2-row calf house under construction

9. A designed feed preparation and storage area.
10. Accommodation for no more than sixty calves in one air space.

In all calf housing designs ensure:

1. That an all-in all-out system is practised with proper disinfecting and cleaning between batches. A minimum period of three weeks between batches should be used to calculate the accommodation required for a given throughput per year.
2. The age range in a group of calves should be narrow, e.g. one to two weeks. This will help to determine the number likely to be housed in one air space.
3. Only calves from similar backgrounds should be grouped together wherever practical.

HOUSING DETAILS

Individual Pens for Calves

Individual pens provide effective separation for each calf. This prevents navel sucking and reduces the spread of disease through facial or other direct contact. If railed pen divisions are used contact is not completely prevented. However, it is more important that calves should be able to see each other than that contact is completely prevented.

Individual pens allow individual feeding to be practised and ensure clear identification of each calf, two useful aids to good stockmanship.

The most economical use of space is usually achieved by a linear arrangement of pens. Depending on the batch size and considerations such as the site and the distance of each pen from the food preparation and storage area, most new designs will have two or four rows of individual pens.

This requires a building of less than 10 m (30 ft) span which can be structurally simple and easy to ventilate well. In converting existing buildings single-row or three-row arrangements may sometimes have to be used.

Minimum sizes for individual pens are shown in figure 4.8. Passages should have a clear width of 1.2 m (4 ft) between rows of pens or 1.0 m (3 ft 3 in) between a row of pens and a side wall.

Individual pens should be constructed so that they can be easily cleaned and disinfected. Although timber might appear to be difficult to clean it is found in practice that demountable pens made up of timber hurdles or sections bound together with baler twine are a very

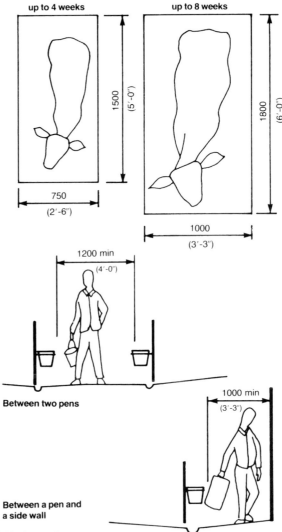

Figure 4.8   Individual pen sizes

satisfactory answer. Figure 4.9 shows an example of such a pen design.

At the end of the rearing period the pens can be completely dismantled and removed. The house can be thoroughly cleaned and the pen sections power washed, dipped in disinfectant and allowed to drain and dry for a few days before being assembled for the next batch of calves.

Alternative materials and methods of construction can be used to meet individual needs and proprietary pens are available. Individual

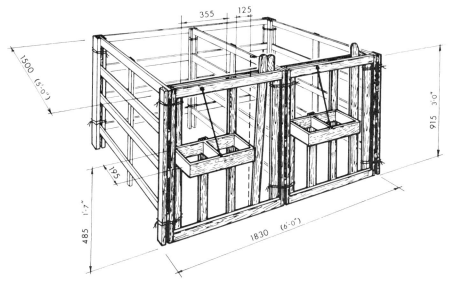

Figure 4.9   Pen design using hardwood timber 20 mm × 50 mm

designs must be judged on their merits asking such questions as the following:

Durability   Will the pen stand up to continuous cycles of use and cleaning, and how easy is it to repair?

Access to calf   Can each pen be opened separately?

Ease of cleaning pen   Can all parts of the pen be easily and effectively cleaned and sterilised?

Flexibility   Can the individual pens be converted into group pens if required?

Price   Do the pens fulfil the above requirements at an economic cost? What is included in the price?

Solid or Open Pen Divisions?

This has been a contentious point for many years and has only recently been sorted out. Today a designer can simply use either. Generally, individual pens should be provided with railed sides. Solid pen divisions need only be used in draughty calf houses where, for some reason, a farmer cannot change the design to reduce draughts, or at the end of rows exposed to cross passages. The advocates of solid pen divisions usually point out that they stop dung passing from one pen to the next; but have they ever tried cleaning them!

Solid divisions are also more expensive, so use them only when necessary.

Because of problems over solid versus open pen divisions it is important to appreciate that a basic misunderstanding was promulgated in the original Brambell Committee Report of 1964.

In the final report it was stated that calf pens should preferably be solid. However, in the draft report this had been applied only to pens for *veal* calves under the old crated veal system before the development of group-reared veal production. It certainly did not apply to normal calf rearing, and the guidance notes stated pen divisions could either be solid or open. However, this option was arbitrarily removed from the published report which stated that solid pens had to be provided—without any evidence or even detailed discussion!

For many years this could not be reversed and was generally adhered to. Calf houses with open railed divisions received grant aid, but if package-deal building companies wished to revise the design then solid pen divisions were required. Farmers then had the extra expense of removing these and providing railed divisions.

Group Pens with Individual Feeding

Calves can be bucket fed individually in group pens holding up to six calves. This offers slightly greater flexibility of layout as the pen footage required is a little less per calf than with individual pens. Space requirement is lower, in that a pen area of 1.5 square metres (16¼ sq ft) can be used for calves up to twelve weeks, and the cost of individual pen divisions is saved. Six calves per pen has been found in practice to be the maximum number that one man can easily cope with when feeding with buckets. The use of yokes for milk feeding can ease the handling of larger groups, but they incur an additional cost and are not recommended in the United Kingdom.

Table 4.3 shows the space allowances and consequent pen depths based on a minimum pen frontage of 350 mm per calf. If the pen frontage is increased as might be the case if, for example, a simple yoke is incorporated in each pen, pen depth can be reduced while the area is maintained.

Table 4.3   Group pens for calves, bucket feeding

Age of calf	Area per calf (m²)	Min. pen frontage per calf (m)	Max depth of pen (m)
Up to eight weeks	1.1	0.35	3.1
Up to 12 weeks	1.5	0.35	4.3

Group Pens with Automatic Feeders

Greatest flexibility of layout is achieved where automatic feeders are used, as the only pen frontage required is for access to the machine and other feeding and drinking equipment, a gate into the pen and a yoke for restraining individual calves. It is important that access to the machine should be good so that the daily cleaning operation is not neglected.

Group size is limited only by the capacity of the feeder machine and the number of teats provided which will vary according to the manufacturer.

The floor space allowance should be a minimum of 1.1 m² per calf up to eight weeks and 1.5 m² per calf up to twelve weeks.

Machines from the various manufacturers vary in dimensions, number of teats and the number of calves they are designed to feed. Machines with only one teat have to be sited one to each pen, but with two or more teats a machine can be placed to serve two or sometimes more pens.

The machine should be placed above or at the side of a drain to ensure that spillage and urine produced near the machine drains quickly out of the pen. Liquid intake is greater with automatic machines than with bucket feeding and properly sloped floors to drain the higher liquid output away are essential.

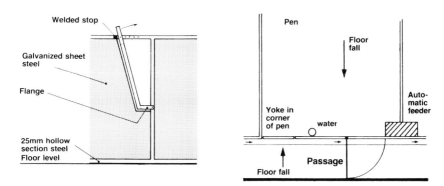

Figure 4.10   Group pen with automatic feeder and yoke

Feeders should be set into the pens to leave the passages clear, and always arrange drainage away from the calves. Water bowls for group pens should be positioned so that any spillage from them also drains away from the bedding but it should preferably be at a distance from automatic feeders to encourage milk consumption.

Recently developed computer-controlled automatic calf feeders allow individual feeding of group-reared calves. This allows simpler group feeding, saves on feed and as capital costs decline the technique should spread. Studies by Dr Welchman at the Bristol Veterinary School also showed reduced health problems with this type of feeder.

The Tethered Feed Fence

The tethered feed fence was developed as an alternative to individual pens for the first few weeks of a calf's life. I worked with raised calf crates—got fed up lifting calves in and out—chopped the legs off the crates then started from scratch looking at the spatial needs of the calf, and finally arrived at the tethered feed fence.

Dung and urine are deposited backwards so the tethered unit is rarely contaminated and as the calf tends to grow upwards and length-ways it remains comfortable. I prefer to group calves as soon as possible so most tethered feed fences are only used from one to three weeks of age. They can be used comfortably up to weaning.

It achieves the objectives of individual penning—prevention of intersucking, prevention of facial contact, separate feeding for each calf—while providing better individual access. It is also slightly cheaper in capital cost than individual pens. It has the additional advantage that it can also be used after the tethered stage as a feed fence and pen front for calves up to twelve weeks of age by simply releasing the calves from the tethers. Alternatively it can be used simply as a nursery stage before grouping calves in a separate building. The space behind the fence forms a group pen and the divisions which prevented contact in front of and behind the fence can be removed if desired for the grouped stage as they are then redundant.

Layouts for the tethered feed fence will be similar to those using

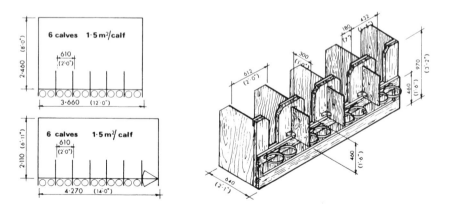

Figure 4.11   Two designs and construction details for the tethered feed fence

individual pens. If the calves are to be grouped behind the fence after the tethered stage, the distance to the back wall should be calculated from the space requirements given for group pens. It would be desirable to include a gate to the passage in each group pen, and the additional frontage required by this will reduce the required depth of pen slightly if the area per calf is maintained. Figure 4.11 shows examples of pens for six calves with and without access from the front.

The detailed dimensions of the tethered feed fence are also shown in figure 4.11. The construction may differ from that shown provided adequate strength is maintained, but the dimensions should be adhered to strictly.

It is fully accepted from a welfare point of view, as the tether is long enough to allow the calf to lie down comfortably and groom itself. It is also now included in Ministry of Agriculture leaflets. A detailed drawing is available from the author.

Floors and Drainage for Straw-bedded Calves

The Floor

The most suitable material currently available for calf house floors is concrete. It is cheap, fairly easily cleaned and can be laid to give the required falls and drainage channels. A damp-proof membrane should always be included in the floor. A wood float finish gives the best compromise between an easily cleaned surface and a nonslip finish for bare concrete areas. A steel float finish in bedded areas will be easier to clean. Slatted and scraped concrete floors have been tried without bedding but are unsatisfactory because the calves can be lying on a cold dirty floor.

Drainage

Good floor drainage is essential in all calf houses. Simply put, excess moisture is removed by good drainage and good ventilation. If the drainage is poor then an excess load is placed on the ventilation system. The calf lies on a damp bed and approximately one-third more bedding is needed. The use of automatic calf feeders means more liquid into the calf and more out at the other end, making drainage even more important than for bucket feeding.

Always arrange floor slopes so that drainage is *away from the calves* and removed from the building as quickly as possible. Also ensure that feed passages are adequately drained.

One problem with calf house drainage is site supervision. Time after time a good drainage layout on drawings has ended up as flat floors either because the builder 'forgot', 'thought he knew better' or simply because site supervision was poor. It is very annoying to have to re-lay floors in order to achieve good drainage in a new building.

Floor Slopes beneath Straw Bedding
For years slopes advised were totally inadequate. As a student I listened to many speakers advising slopes which simply do not allow any drainage from beneath straw bedding in calf pens. The drawings for one architect-designed calf house in Scotland even had a slope of 1 inch in 18 ft to go beneath straw bedding! This is very difficult to set up and resulted in wet bedding and a poor environment. Over several years I tried different floor slopes beneath calves and quickly found than 1 in 20 is the best practical solution. If this is taken to extremes, such as 1 in 10, good drainage occurs, but straw bedding, the calves and anything else mobile ends up at the bottom of the slope!

If the bedding is allowed to build up excessively and compresses, then drainage will not occur, but with up to 23 cm (9 ins) of straw floor slopes of 1 in 20 are adequate.

On a long floor a slope of 1 in 20 may mean a considerable rise. To avoid this, for instance in a group-reared monopitch scheme, consider splitting the long slope into two.

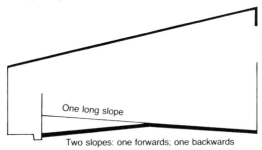

One long slope

Two slopes: one forwards; one backwards

Floor Slopes for Passages
Channels at the side of passages should have no straw in them so a very shallow slope will ensure liquids flow away.

For individual pen layouts, drainage channels should run under the feed and water buckets at the pen fronts and drain both pens and passages (see figure 4.12).

If channels are at the back of pens, spillages from the feed and water and in the passage will have to run through the bedding and will then pass through other pens.

The passage should be domed or sloped towards the drainage channel to keep it clean and dry. On no account should the drainage channel be placed in the centre of the feeding passage. It will result in

Good practice

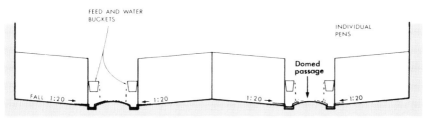

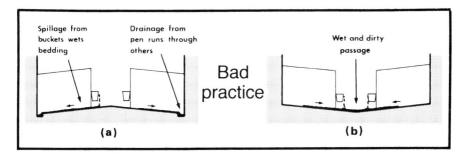

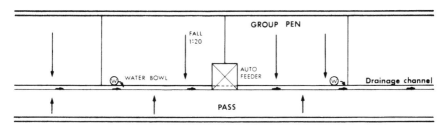

Good practice

Figure 4.12   Drainage layouts

the passage itself being unpleasantly wet and dirty. Who wants to work in an open gutter!

In group pens with automatic feeders the floor falls within the pens should be towards the feeder and water bowl, so that spillage and urine are quickly led away from the bedding. For convenience, set feeders into pens with railed sides to protect them. Placing automatic feeders in passageways hinders easy movement along the passages.

Duckboards are sometimes used in individual pens to improve the drainage on level or near-level floors. This is a desperate measure for emergencies only. A considerable amount of labour is, however, required to clean them, and rats can live in the space between the slats

Plate 21   Good natural and artificial lighting

and the floor. Re-laying the floor to the correct falls is the best solution
to the problem of poor drainage.

Lighting

Good artificial and natural lighting should be provided to enable calves
to be observed easily at all times.

Lighting does not appear to have any major effect on the physiology
or behaviour of housed calves. Adequate lighting must be available for
the stockman to carry out the routine tasks that take place in the
building, including inspection of the calves. This could be provided by
openings, glazed or unglazed, equal to 10 per cent of the floor area. To
satisfy the requirement by artificial light 100 lux should be provided
at floor level. In practical terms this would be about 20 W of tungsten-
filament lighting for each square metre of floor area, or approximately
one-third of this power input if fluorescent lighting is used.

The only problem with fluorescent lighting in farm buildings is at
low temperatures and a technical note entitled 'Fluorescent light for
low ambient temperatures' is available to overcome this problem from

the Electricity Association's Farm Electric Centre, NAC, Stoneleigh, Warwickshire. They also produce technical information sheets on the design of natural and artificial lighting for farm buildings and a handbook 'Essentials of farm lighting'. These publications are all free of charge.

Selecting Materials for Walls, Pen Divisions and Roofs

Internal wall surfaces at calf level and pen divisions should be able to withstand the calf's tendency to lick, chew, kick and defecate against anything within reach. Among commonly used materials are cement-plastered blockwork or brickwork, steel rails, plywood, softwood, hardwood, fully-compressed asbestos cement sheet, and oil-tempered hardboard.

Timber can be difficult to clean thoroughly, especially if it is not dressed. Tests in Sweden have shown that some hard, smooth materials such as galvanised steel, smooth surfaces painted with alkyd PVA or epoxy paints and laminated plastics can be surprisingly difficult to clean, so the choice from this point of view is not simple.

Calves will also chew timber, but this can be prevented by treating the timber with creosote after it is cleaned. The creosote should be allowed to dry out for a few days before calves are put in, otherwise their skins may suffer burning.

It is important to avoid all risk of poisoning from lead-based paint, which should never be used on agricultural buildings. Particular care is required with old painted items such as reused doors or other second-hand timber.

Soft materials such as polystyrene or polyurethane insulation, soft building boards, or asbestos substitute insulation boards are very quickly damaged by calves. Remember that a calf can reach a considerable distance up a wall. Even when used above calf level these materials may be damaged when using a high-pressure hose for cleaning so they are best avoided.

Disposable materials sometimes have a place as walls or pen divisions in calf houses. The most common material of this type is baled straw. The number of bales required and the space they take up is excessive for individual pens except in makeshift circumstances with small numbers of calves. Roof materials and structure should be such as to permit periodic cleaning and should not harbour vermin.

Food Storage and Preparation Area

Storage and working space is required within or adjoining the calf house for efficient operation. The minimum provision is a large sink

Plate 22   Feed storage and preparation area incorporating an Economy 7 electric water heater

with hot and cold water supplies, good lighting, power sockets, storage space for milk substitute (off the floor) and concentrate feeds, and shelves and cupboards for veterinary supplies. The floor should be easy to clean with a fall towards a drain near the sink. This will prevent any spillages seeping under feedstuffs. Sufficient clear working area for measuring and mixing feeds, manoeuvring trolleys or barrows and bringing in and stacking feed bags is essential.

Hard-and-fast rules for the size of storage and working area cannot be given because of variation in feeds, feeding method, housing period and the shape of the room, as well as the location of nearby storage areas. Frequent replenishing of a very small store from a distance, or the managing of a house with no nearby storage will be time-consuming and tedious. Whilst it might be thought ideal to have space to store all the feed required for rearing a complete batch of calves until they leave the building, this may be excessive, particularly for larger houses and where monthly feed deliveries can be made to the door by the supplier.

The area should have wide doors to the outside for taking in feed and supplies, and also have good access into the calf accommodation. The floor should have a fall of 1:40 towards a drain which should be beneath the sink. The sink itself should be provided with a large draining board for equipment and utensils to drain after washing up. All internal surfaces should be smooth and easily cleaned, and care should be taken to make the room vermin-proof and free from crevices which could harbour any vermin which gain access through doors left open. A cupboard will be required for storing veterinary supplies.

ECONOMY 7 WATER HEATING

The most economical way to heat water electrically for calf feeding is to heat it during the night on a cheap night rate tariff and store the hot water until it is required for use. It can reduce the cost of warm milk feeding for instance from £1 to 40p per calf reared.

All electricity distribution companies offer a tariff for farmers which gives 7 hours at a cheap rate during the night-time (usually in the period between midnight and 0800 GMT). This is most commonly referred to as the E7 tariff.

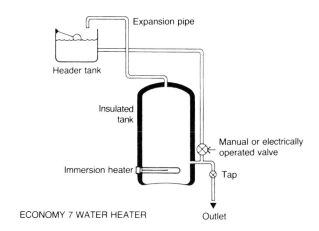

ECONOMY 7 WATER HEATER

Features of a Good Economy 7 Water Heating System

If maximum benefit is to be made of the cheap rate period of the E7 tariff for water heating for calves, then the following points will be important:

Tank capacity The tank, or tanks must be adequately sized, so that all water can be heated and stored during the cheap night rate period.

Insulation   As the water may have to be stored for 12 hours or more, good tank insulation is essential, so that a minimum amount of heat is lost from the tank surface. A minimum insulation level of 0.45 W/m² deg C (equivalent to 50 mm sprayed polyurethane) is recommended.

Timeswitch   The immersion heater must be switched on in the E7 cheap rate period by using a locally placed timeswitch.

Design of the Tank and Pipework System

A system is recommended where the tank is both filled and drained from the bottom. The action of draining is done by gravity after the cold water inlet valve has been closed. This should provide all the hot water for calf feeding on the E7 tariff. The immersion heater can be switched on for a short time on the day tariff if for any reason more hot water is needed. Note that the immersion heater is installed below the level of the water outlet. This should ensure that, if the heater is accidentally switched on when the tank is drained down, there is less risk of damage to the element.

A manually operated valve can be used on the cold water supply; alternatively a solenoid valve can be installed to open automatically at the beginning of the E7 cheap rate period. This eliminates the risk of the stockman forgetting to refill the tank after it has been drained down.

All exposed pipe work close to the water heater should be insulated and protected with electric pipe trace heating to eliminate the risk of pipes bursting under winter conditions.

BASIC DESIGN DATA FOR CALVES UP TO
TWELVE WEEKS OF AGE

Environmental Requirements

1. Minimum cubic air capacity (air volume) when fully stocked 6 m³/calf (212 ft³/calf).
2. Ventilation. With eaves level air inlets and ridge outlets allow an inlet area of 0.045 m²/calf (0.5 ft²/calf), an outlet area of 0.04 m²/calf (0.4 ft²/calf) and 1.5–2.5 m (5–8 ft) height difference between the two. If pens are arranged along the side of a naturally ventilated building, then a cover should be placed over the rear part of the pen to prevent incoming air from dropping onto the calves. If fans have to be used design on a basis of 35–105 m³/h (20–62ft³/min) per calf.

3. Air movement rate close to the calf, not more than 0.25 m/s (50 ft/min) in winter.
4. Relative humidity and temperature. No specific requirements under UK conditions. Satisfactory if similar to outside conditions.
5. Insulation is not required in monopitch or pitched roof designs.
6. Heating. Localised electric heating may be required for sick calves. Use dull emitter bulbs or panel heaters. For group heating of sick calves consider quartz linear lamps.
7. Lighting. Natural—provide 10% of the roof sheeting as translucent sheeting. Artificial—provide 100–200 lux at calf level to aid inspection of calves.

Passage Width

Two rows of pens—one on each side of a central passage 1.2 m (4 ft).
    Single row of pens on one side of a passage 1 m (3 ft 3 in).

Water

Individually penned—one water bucket for each pen.
    Group penned—one water bowl for every 10–12 calves.
    Water heating—provide enough capacity to run on night rate Economy 7 electricity tariffs. Provide pipe trace heating and insulation to prevent freezing of water pipes.

Trough Frontage

Feeding space for individually fed calves—350 mm (1 ft 2 in) per calf.

Floor Design

Beneath straw bedding provide concrete floors with a slope of at least 1 in 20. Passages should be domed and the floor slope in feed storage areas should be 1 in 40.

Table 4.4 Pen space requirements

	Age	Minimum space required per calf	
Individual pens	Up to 4 weeks	1.1 m²	(1.5 × 0.75 m)
		12.5 ft²	(5 ft × 2 ft 6 in)
	Up to 8 weeks	1.8 m²	(1.8 × 1.0 m)
		19.5 ft²	(6 ft × 3 ft 3 in)
Group pens	Up to 8 weeks	1.1 m²	(12.5 ft²)
	Up to 12 weeks	1.5 m²	(16.5 ft²)

Design Checks

Ensure that the following points have been taken care of in your final design. The most common faults occur in these areas.

Drainage

Are floor slopes adequate in

- pens?
- passages?
- feed storage and preparation area?

Are liquids let out of the building

- without fouling working areas?
- without passing through calf pens?

Electrical Details

The Farm Energy Centre (part of EA Technology) at the NAC Stoneleigh and area electricity board agricultural engineers provide a free energy efficiency service and free publications, tel. no. (01203) 696512; fax (01203) 696360.

Ventilation

Natural Ventilation
Is the total air inlet area adequate?
Is the total air outlet area adequate?
Are the inlets well distributed?
Will the calves be free from draughts?

Air speeds at calf level can be checked with a hot-wire anemometer from: Airflow Developments Ltd, Lancaster Road, High Wycombe, Bucks HP12 3QD

Ventilation patterns can be checked with smoke pellets or smoke tubes. Smoke pellets are available from: Thermal Products, Warfield Works, Glenview Road, Eldwick, Nr. Bingley, Yorkshire. Smoke tubes are available from: Mine Safety Appliances Co. Ltd, East Shawhead, Coatbridge, Lanarkshire ML5 4DT.

Mechanical Ventilation
Used only when natural ventilation is not possible in some conversions. Use manual *not* thermostatic fan controls.

Table 4.5 **Heat production of cattle confined at thermoneutrality and at different stages of production**

	Body weight (kg)	Surface area (m²)	Heat production			
			MJ day⁻¹	MJ m⁻² day⁻¹	W m⁻²	W
Fasting metabolism						
Calf 1 month old	50	1.24	9.4	7.6	88	109
1 year old	300	4.11	33.2	8.1	93	384
Steer 2 years old	450	5.39	34.7	6.4	74	401
Growing Cattle						
Veal calf 1.5 kg gain/day	100	1.97	26.3	13.3	154	304
Baby veal 1.0 kg gain/day	150	2.58	31.0	12.0	139	359
	350	4.55	56.7	12.5	144	656
1.3 kg gain/day	150	2.58	33.5	13.0	150	387
	350	4.55	61.1	13.4	155	707
Store cattle maintenance	250	3.64	38.7	10.6	123	448
0.4 kg gain/day	250	3.64	49.8	13.7	158	576
Fat stock 0.8 kg gain/day	450	5.39	73.0	13.5	157	845
1.5 kg gain/day	450	5.39	81.5	15.1	175	943
Beef cow maintenance	450	5.39	50.0	9.3	107	578
Dairy cattle						
Dry, pregnant	500	5.79	52.2	9.0	104	604
2 gallons/day	500	5.79	64.6	11.1	129	747
5 gallons/day	500	5.79	77.0	13.3	154	891
8 gallons/day	500	5.79	89.4	15.4	178	1,034

Basic Thermal Data.

Is fan capacity adequate at the working back pressure? Is the air inlet area adequate? Are the inlets well distributed? Will the controls ensure that the ventilation rate will not go below the minimum required?

Will the calves be free from draughts? If an existing calf house has fan ventilation with thermostatic control change to manual speed control, ensure that the fans run continuously and check that the fan output is adequate.

Materials Handling

Is the size of the feed storage and preparation area adequate?
Is it positioned well in relation to the calf pens?
Can bedding and muck be easily brought in and removed from the building?

Chapter 5

CALF HEALTH

THIS CHAPTER on health and disease of the calf is a simple outline of sensible approaches to minimising the risk and spread of disease. It is not meant as a treatise on calf diseases or disease treatments, merely a series of pointers to enable those rearing calves to understand basic attitudes and procedures which will maintain good health and productivity.

Co-operation between the Farmer and his Veterinary Surgeon

Good calf health is basically the result of teamwork involving the farm staff and any source of professional help that is available. It is very important that the veterinary surgeon knows his farmer customer's calf-rearing facilities and his broad policy, whether on a home-bred unit or a specialist unit for buying in calves. A meeting on site is obviously the best place for discussion and the exchange of ideas. Quite often this can take place at a suitable time to consider results and problems in the past, or to look forward to a change in the enterprise, particularly where expansion is planned with new or modified buildings. The impact of such things as changes in feeding methods, sources of calves and new staff on the health of the calf are important topics.

A great advantage of setting up an advisory visit of this type is that the veterinary surgeon views the rearing enterprise with a fresh eye, and one which is both professional and experienced. This is especially true in larger veterinary practices where there can be a degree of specialisation within the team. This specialisation is being increasingly backed up by modern facilities for diagnosis and treatment with computerised record keeping both for the animal records and for the accounts.

This latter feature is to be very much welcomed by the rearer who can be told the cost of advisory or emergency visits and the cost of drugs, and then see these itemised on his statement.

Support Services
The veterinary surgeon can also describe his access to further in-depth advice and facilities. Probably most important are the Veterinary Investigation Centres of MAFF (or the Department of Scotland V.I. Centres). Here, as well as statutory work in connection with notifiable diseases, specialist diagnostic work can be carried out, particularly on the bacteriology front. These Ministry Veterinary Centres in turn are further supported by the Central Veterinary Laboratory at Weybridge, and by University facilities with their field stations.

On bought-in calf units there is a constant risk of importing infections. When illness flares up the veterinary surgeon will need rapid diagnostic help to identify and type bacterial organisms and their susceptibility to the different antibiotics from samples taken before and during any treatment. A good example of such investigation is the scouring calf. It is often necessary to determine the susceptibility of the bacteria causing scour (enteritis) to the available antibiotics. Virus infections are much more difficult to diagnose, since they cannot be isolated and cultured so readily in the laboratory. In the short term they have to be diagnosed by the experience of the veterinary surgeon. In the longer term a good deal of help can be given by the laboratory by diagnosing from successive blood samples showing a rising immunity (antibody titre) against the infection. Positive isolation or identification of viruses is sometimes possible from samples taken from calf faeces, blood or the mucous obtained on nasal and ocular swabs.

Lastly, the veterinary surgeon is able to assess the skills of the stockperson in charge of the calves and to guide him in treatment and nursing practices. In particular he would stress the need for keeping simple records of illness and treatments, as well as the proper use and storage of drugs.

Signs of Ill Health

There is no doubt that the successful rearer is one who can recognise ill health very early—almost before the calf has shown the classic signs of a particular problem.

The good stockperson must have acute observational powers and additionally must have the time available and the facilities of space and light to be able to exercise those powers.

The appearance and activities and any changes from established behaviour are the first things to note. If the calf is lying down when it should be up, or if it lies down quickly after the routine disturbance of feeding or bedding down, this should be noted.

The posture in the lying position should be assessed. When

ruminating, the calf usually sits on its haunches with the two front legs tucked underneath its chest. If the front legs are stretched out in front, this could indicate an attempt to relieve pressure from chest pain arising from pneumonia. When seriously ill the calf may lie completely flat and be unable to lift its head. When up on its feet, the calf's gait and type of movement can indicate lameness or injury. Stiffness, moving in circles or sheer weakness would point more to infection.

Such a calf should have been recognised as being ill long before this stage however, and should not therefore be confused with one having a sound sleep on a full stomach especially noticeable on the ad-lib systems of feeding.

The eyes are a good indicator of health. They should be prominent and bright, not sunken or glazed, and there should be no copious tears and discharge.

Blindness in one or both eyes can of course occur and should be taken as a sign of serious disease. The calf's ability to see should always be checked.

The calf's ears, similarly to the eyes, are a good indicator of good or bad health. One or both may drop from their customary pricked-up position, and at the same time become cold to the touch. The nose should be clean and damp (not dry) and with no frothy or thick mucous coming from the nostrils.

The rate of respiration of a calf normally lies between 20 and 40 per minute when at rest and should be calm and regular. Obviously activity, even within a small pen, will raise this temporarily as will high temperatures in the house.

Similarly when the calf is ruminating after an intake of feed its breathing rate will often rise, and this is entirely normal.

The abdomen is likely to be empty and 'tucked up' if the calf is not feeding. Perhaps a better indication therefore of chest infections is the rate combined with the shallowness of breathing, which may be accompanied by grunting, whistling or coughing.

The condition of bloat where the abdomen is distended by gases is well known to stockpersons. It is usually more pronounced on the calf's left side and may be severe enough to require immediate attention; it may however be only a slight distension. Calves which have persistent mild bloat fail to thrive.

The calf's tail should be clean and active, not slack and lacking tone, nor should it be stiff. A clean tail is of course an indication of normal consistency of the faeces, the latter being firm, but without being so dry as to cause discomfort to the calf when passed.

Looseness, or scouring, may occur and become progressively more marked until the faeces are quite watery. This scouring condition is obviously far from normal. Note that very watery scour may not be observed early if the liquid is absorbed by a straw bed and the tail and rear end of the calf are not wet and stained.

The navel of a newborn calf provides an easy entry for infection. It should be checked for swelling and hardness which may be with or without pain; expert advice is necessary to differentiate between a septic navel or a rupture in this region.

When considering any area affected by injury or local infection, the basic signs of heat, pain, swelling or discoloration, and discharge, are all valuable indicators of the nature and importance of the problem.

The skin and hair of the calf are excellent indicators of its state of health. The hair should generally lie down and be smooth and shiny. However note that the hair type varies a great deal both between and within breeds in the degree of curliness and roughness. The tightness of the skin is a useful indicator of the degree of dehydration of a calf.

After weaning, calves which are on ad-lib feed and thriving exceptionally well with a large feed intake tend to sweat more. This is obvious if building ventilation rates have not been increased to match the much greater air requirements of the growing stock.

Loss of hair by rubbing is indicative of infestation by lice. Careful examination will reveal the lice eggs laid on the hairs as well as the lice themselves. The routine dressing of all the calves in the group with louse powder repeated two weeks later should be part of the normal routine.

Plate 23   Lice eggs on hair shaft

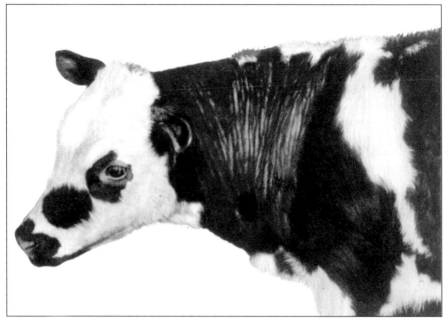

Plate 24 Lice—coat tends to be arranged in lines along the calf's neck

Part of the skill of observation of a group of animals (whether loose housed or in single pens) is to watch more closely those calves which have contact with diseased stock, even though they are apparently healthy. At the same time, if it is at all possible, the sick animals within a loose-housed group stand a better chance of recovery if they can be isolated in an area where they can be more readily observed and nursed. This isolation will remove a potential source of infection from the other calves. For single-penned calves it is important to make a thorough cleaning break once the batch is through.

It is extremely useful to check the calf's temperature as a confirmation of suspected infection. It will sometimes show acute infection by a very high reading (say 105–106 °F; 40.5–41 °C) in a calf which is only just beginning to look off colour. Generally if the calf's temperature is 102.8 °F (39.3 °C) or over it indicates an infectious disease.

When antibiotic treatment is undertaken the calf's response can be monitored by further temperature readings, until normality is achieved (101.5–102.8 °F; 38.6–39.3 °C). The temperature of the calf is taken in the rectum with the thermometer pressed against the side of the passage for a minute. It should be wiped clean and read carefully, and of course shaken down before another reading is taken.

The Calf's Defensive Mechanisms

The ability of the calf to withstand infection and adverse conditions varies a great deal. Even in a severe outbreak of disease, such as viral pneumonia or salmonella, not all animals within a group will succumb. Resistance to infection in the form of circulating antibodies is initially passed to the calf by its dam in the colostrum. Further antibodies (immunity) may later develop in response to an infective agent. The importance of the calf receiving an adequate amount of colostrum in the first few hours after birth is well known among stockpersons. When calves are reared at home, then there is every incentive to see that the calf has been well fed with colostrum soon after birth. However, the biggest problem lies with purchased calves. Surveys have shown that many of these calves have received inadequate amounts of colostrum and that there is a very strong correlation with subsequent illness in these calves. There are now methods of quickly checking antibody levels in the blood of the calf which enable the rearer to be confident that the correct care has been taken in the feeding of the calf. As yet such checking is not widely practised.

Spare colostrum need not be thrown away. It can be left to go sour in a suitable container at room temperature, and then fed over the next fortnight. Although antibodies cannot be absorbed into the system after the first day, the colostrum is still a beneficial food, though the system is rather 'messy' for most farmers.

There is mounting pressure for legislation to prevent the movement of young calves from market to market too frequently, and to ensure that only healthy calves which are sufficiently old to withstand transporting and penning are presented to the market.

Since January 1st 1985 there has been a provisional code of practice drawn up by the Livestock Auctioneers Market Committee, following discussions with the British Veterinary Association, the Royal Society for the Prevention of Cruelty to Animals and the NFU.

The main points are:

1. Auctioneers should actively encourage the feeding of colostrum to all calves before they are entered for sale.
2. Any calf less than fifty-six days old should not be accepted for sale in a market if the auctioneer has reason to believe that it has been presented for sale in more than one other market during the previous twenty-eight days.
3. Auctioneers should refuse to accept for sale any calf which is less than seven days old or which has a wet navel.
4. Auctioneers should ensure that before any calf is accepted for sale, purchasers can learn the vendor's true name.

5. Calf pens should not be overcrowded; adequate clean bedding should be provided. Calves held in any one pen should be of an even size and no calf should be tied by the neck while it is in the market.
6. Once a calf has entered a market it should not be moved unnecessarily. Calves should not be lifted by the tail, neck or ears, and the use of sticks or goads should be banned. Particular care should be taken when loading or unloading calves.
7. Any injured, sick or distressed calf should not be allowed to enter the market. It should instead be taken to an isolation pen for immediate veterinary attention.
8. Calves should be protected from excessively cold or hot conditions and, where necessary, proper feeding and watering arrangements should be made.
9. The timing of calf sales should be reviewed regularly with the aim of minimising the time that any calf spends in the market or in transit.
10. Freemartins (i.e. most heifers born twin to a bull) should be declared.

The resistance to infection conferred by the antibodies is of course not absolute and may not offer protection against organisms or strains of organisms different from those on the farm of origin, or indeed against massive levels of infection.

The immunity from the colostrum is a passive one and declines with time and has to be replaced by the calf's own natural build-up of active immunity. This will take place best in an environment which does not present too early or severe an infectious challenge. Good husbandry in the form of suitable housing and feeding and avoidance of too many changes are also important. The avoidance of stress is a keynote and includes factors like chilling by wind and rain, irregular feeding, inadequate bedding, being trampled within the group, or even lack of rest. Most of these factors are easily avoided in home-reared calves; it is the purchased calf which is most at risk.

In order to boost the calf's defensive mechanisms as quickly as possible sera or vaccines (or a combination of both) may be used. These products are usually administered at intervals to give proper immunisation either by injection or as nasal droplets.

An antiserum is a fluid containing antibodies which actively destroy bacteria and viruses to which they are specific. They may be found in cows' colostrum or as commercially available antisera preparations. The latter, such as the serum for tetanus, give immediate protection against infection.

A vaccine is a substance which stimulates the production of antibodies against a disease organism. Vaccines require a period of time (often with repeated doses being necessary) in order to produce the required immunity. Vaccines and sera vary in their efficiency for disease prevention and treatment. For example the vaccine against the organism causing tetanus is very reliable whereas those vaccines used to combat E. coli infection (a cause of enteritis in calves) sometimes fail in their effectiveness, due to the many variants of E. coli which cause disease.

A unique vaccine is the one to control husk caused by a nematode (worm) in the lungs. It is given orally before young cattle are turned out to grass. Two doses are needed at a minimum of six and two weeks prior to the turnout. Unless severely challenged, cattle in subsequent grazing years have sufficient immunity to resist the infection even though pastures are carrying infective larvae.

Vaccines are available for one or two of the species of salmonella, many variants of E. coli, as well as several of the viral and bacterial organisms which go to make up the complex of viral pneumonia.

Alternatively a course of prophylactic treatment with an antibiotic administered daily for a short period of about five days in the liquid feed can be given. This preventative measure is usually only practised on bought-in calves after advice. Once the calves are older and ruminating it is not possible to use antibiotics by mouth without interfering with rumen function. This treatment used properly does give a group of calves a chance to overcome the stress and possible infections of the market. Oral feeding of antibiotics has potential dangers and should only be undertaken when absolutely necessary and only then on expert advice.

Hygiene

On the veterinary surgeon's advisory visit the whole calf unit needs to be assessed to see how the calves, people, food and water, urine and solid manure are moved within the buildings or yards. As far as possible the crossing of pathways should be avoided. In particular, the movement of contaminated material should be by the shortest and quickest route to a safe place to prevent it being a further hazard.

Feed should be used sequentially to avoid staleness, and kept in dry areas with active steps taken to keep it free of vermin. Adequate clean water is an obvious necessity and individual buckets or self-filling water bowls are preferable to open tanks. Water bowls need daily checking to see that they are functioning correctly and have not been fouled. Good siting and protection are important for a proper supply to be maintained in frosty weather. Mouldy hay or straw can cause respiratory problems and only clean dry materials should be used.

As yet calf units are not subject to the same stringent rules about visitors and lorries as are pig and poultry units, but as much 'boot hygiene' and the use of separate staff for sick and healthy groups should be practised as is possible, especially where salmonella infections are involved.

Proper washing facilities for the staff on the site helps promote good hygienic practices.

However the principle of regular de-stocking and cleaning of the houses is the most important factor in breaking the cycle of infection in calf houses. The modern tendency to use removable wooden penning, preferably sited clear of the walls, enables rapid mechanical cleaning of houses followed by pressure washing and disinfection.

Finally, whoever is ultimately responsible for the success of the calf unit must ensure that the facilities are used correctly and the rules discussed above are adhered to at all times. Over time, procedures often become lax. This must lead to failure.

Disease Organisms

Because micro-organisms are so small and exist in such large numbers they are very easily transmitted by animal and human movements, on equipment and in the air. New infections can arise from vermin, contaminated feeding stuffs or on the wheels of lorries.

However, the calves themselves are the most important source of infection with both bacteria and viruses being passed from one to another with the complication that the donor animal may not be sick, but apparently healthy, i.e. just a carrier. Some organisms, notably the bacterium E. coli, are always present in the gut of healthy calves but can cause disease when given the right conditions to be virulent. Similarly a calf whose resistance has been lowered by cold and wet and poor feeding, will contract pneumonia caused by a virus apparently from nowhere.

On the other hand some infections can be categorically said to be absent from the farm. They are so virulent that stock would succumb if they were present. Perhaps foot and mouth disease, tuberculosis and brucellosis would be the best known examples. Others like anthrax or tetanus can lie around in the soil and buildings in a protected form for years just waiting to strike. Some of the other most common organisms likely to cause problems are shown in table 5.1.

All these living organisms multiply in or on the animal body and cause disease to a varying degree of severity. This will depend on the

Table 5.1   Common disease organisms

Disease	Disease Organism
Calf scour	K99+ E. coli
	Rotaviruses
	Salmonella
	Cryptosporidia
	Bovine Viral Diarrhoea
Pneumonia	Pasteurella
	Infectious Bovine Rhinitis
	Parainfluenza type 3
	Adenovirus
	Bovine Viral Diarrhoea
	Mycoplasmas
	Respiratory Syncytial Virus
Navel illness	*Streptococci* ⎫ species
	Staphylocci ⎭
	Corynebacteria pyogenes
Calf diphtheria	Fusiformis necrophorus

susceptibility of the stock, their environment, nutrient and antibody status to the particular strain of invading organisms.

An infected navel, with or without joint ill (or joint evil) is an example of a condition due to bacterial disease. This may start as a small degree of infection at the navel of a newly born calf and eventually progress to severe disease of the navel and adjacent areas. Septic arthritis of the limb joints of the calf may follow and this condition is often progressive and fatal. This is an excellent example of a situation which is easy to prevent by correct care and attention soon after birth, but difficult and often impossible to treat successfully after the disease has become severely evident. Navels should be dressed at birth with a strong iodine solution, or antibiotic aerosol, especially on units where there is a known high risk of infection. An appropriate antibiotic injection may also sometimes be necessary.

Under modern farming conditions disease spread is made easier by the high stocking rates in buildings forced on us by economic necessity; by inadequate resting of houses and the movement of stock between and within farms. This often leads to a sudden severe challenge before the calves have had chance to build their own immunity when the challenge has been low.

One of the characteristics of both bacteria and viruses is their ability to change their identity slightly, producing new strains of increasing pathogenicity which overcome the existing resistance of the animal,

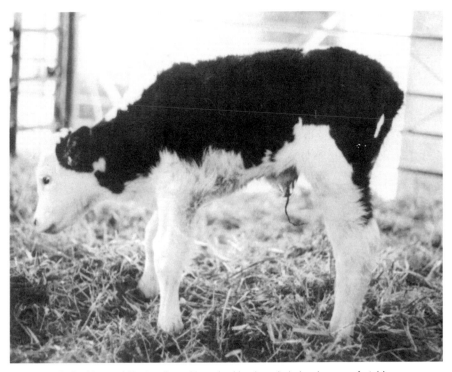

Plate 25   Calf with navel ill, standing with arched back and obviously uncomfortable

and also fail to respond to a particular drug which was formerly effective. Even more of a worry in the long term is the known ability of a relatively harmless organism to acquire resistance to an antibiotic and then to transfer that resistance to another more potent unrelated organism.

Amongst calves both E. coli and salmonella are notorious in having literally thousands of strains and when they are causing a problem on a particular farm the first job is to test their susceptibility to a range of antibiotics. This is easily done in the laboratory with a swab from an untreated calf enabling the correct drug to be found. The veterinary surgeon can then decide dosages and length of treatment to give the most effective and economic therapy without waste of time.

It should particularly be noted that disease conditions with similar clinical signs may have a variety of causes:

- tetanus
- Cerebro Cortico Necrosis (CCN)
- injury to head or spine
- meningitis

Plate 26   Calf with joint ill. Note the swollen front knees

- heat stroke
- abscess in central nervous system
- Bovine Viral Diarrhoea
- Infectious Bovine Rhinitis
- inherited nervous disorders
- vitamin A deficiency
- overdosing with furazolidone
- lead poisoning

Other Causes of Ill Health

Conditions other than those caused by bacteria or viruses are common amongst calves. 'Unthriftiness' in the farmer's eyes can be due to a multiplicity of causes. For example:

- cleft palate—a genetic defect
- selenium deficiency and copper deficiency are nutritional shortages
- blindness from disease or injury

Plate 27   Ringworm infestation, typically on the head and neck

- deprived from food and drink by stronger companions—poor management
- haemorrhage from navel
- stomach ulcer
- impaction of stomach with fibre
- internal abscess from navel
- heart disease
- chronic bloat

Other common conditions occur on most farms from time to time—some reduce thriving and good appearance; others can be fatal.

External Parasites   Lice are the main problem causing severe irritation to the calf which rubs the affected parts causing loss of hair, commonly over the shoulders and round the tail head.

Close examination reveals the lice and also their eggs which are attached to hair.

Dressing with louse powder has to be repeated at ten-day intervals to

Plate 28 Typical 'nose up' position of calf, blind from CCN. It made a full recovery following treatment with thiamine (vitamin B_1).

break the breeding cycle. In summer more persistent chemicals may be sprayed on the calves for fly control, and these will also control lice.

Ringworm Caused by several fungal species, ringworm is more troublesome than lice. It produces typical round areas of hair loss which may be isolated and scattered, or profuse and coalescing into each other causing large unsightly lesions. These areas may become encrusted and sore, and although in healthy animals they will eventually heal and regrow hair, this is a very slow process. Badly affected animals do not seem to thrive and are not liked for obvious reasons in a group of animals ready for sale.

Although hygiene and resting of equipment is recommended to control all infectious agents (i.e. the ringworm fungus) it is unfortunately wishful thinking to say that a house, particularly one with rough walling, is clear of all infection.

For ringworm the best approach is to treat calves as soon as any infection is seen with the in-feed antibiotic, griseofulvin, which is specific to the fungus. A seven-day course of treatment is necessary and the dose is weight related. Treatment is therefore much cheaper and effective if done early, especially if the calves are singly penned or in small groups. A suitable ringworm dressing can also be used to aid a cure of the infection and reduce spread of this disease.

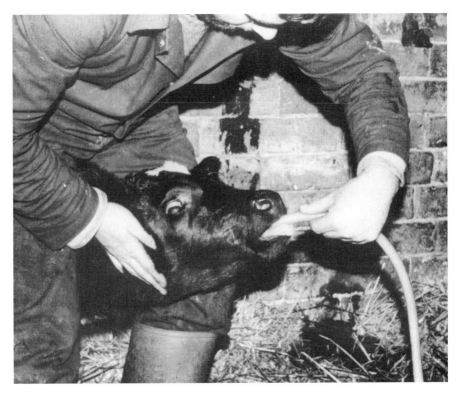

Plate 29 Passing a stomach tube

Cerebrocortical Necrosis   Better known as CCN, this is usually found in young rapidly growing stock where there is a lack of or extremely low level of the B group of vitamins particularly thiamin (B1).

Symptoms are often sudden in onset and are those of trembling or a fit, with varying degrees of disturbance of the gait. Early cases will respond to injections of the vitamin B1.

Lead Poisoning   This disease has similar characteristics to CCN. It is the commonest form of poisoning in livestock and is often seen following the ingestion of paint (however recent or old) which contains lead. Signs are blindness, difficulty in moving and sometimes loud bellowing.

Hypomagnesaemia   This describes a condition where the blood level of magnesium is extremely low. This usually only occurs in calves which are single suckled.

Bloat This condition causes distension of an organ within the abdomen by gases. The degree of distension varies but is quite often progressive to a state of prostration and illness. The usual organ affected is the developing rumen (on the left-hand side of the calf) but sometimes the abdomen or true stomach (on the right-hand side) may become distended with gas.

A simple remedy is to pass a non-abrasive flexible tube via the mouth down into the stomach in order to release the trapped gases. If this does not quickly relieve the problem it is advisable to call professional help. It is important to determine the reason why diseases (which are of a non-infectious nature) may occur. Quite often over-feeding or irregular feeding is the reason for bloat. Occasionally obstruction of the oesophageal passage by a foreign body may cause this condition.

With all of them, careful and continual observation by the stock-person is needed to discover the calf which is abnormal in some way, often masked by others in the group which are perfectly healthy. Early detection is the key to successful treatment.

Lameness in calves can occur from an early age when housed, or later at grass. Again, careful diagnosis is needed to pinpoint the cause. For example:

- accident and trauma to limbs
- spinal injury
- foul in foot infection
- thrombosis in limb following navel infection
- use of dirty needle to inject in rump
- inherited defect in structure of limbs
- tetanus infection

Similarly, eyes can suffer from:

- inherited defects
- New Forest disease—common at grass in hot dusty weather
- pasteurella and IBR infections
- defects following BVD infection while in the womb

Skin is a good indicator of health. Common problems which detract from the bloom on a calf are as follows:
- ringworm—a fungal infection
- mange and lice from external parasites
- burning from heatlamp
- loss of hair around muzzle—fat of milk replacer poorly mixed
- general loss of hair—maybe heredity or shock after severetion
- zinc deficiency

Infections Contracted but Not Apparent Until Later Life

- Enzootic Bovine Leucosis
- Johne's Disease
- Bovine Viral Diarrhoea
- internal abscess
- tuberculosis
- leptospirosis

Factors Causing Sudden Death

- ruptured stomach
- bloat
- twisted intestine
- internal haemorrhage
- lead poisoning
- blow to head
- low blood magnesium (usually only suckled calves)
- clostridial enterotoxaemia
- septicaemia from salmonella or E. coli

Factors Which Affect the Newborn Calf

The stockperson has to attend to the calf from its birth—in fact, quite often is involved in its delivery. We are all familiar with the fact that calves are lost at birth, not just through malpresentation, but quite often due to relative oversize of the calf. Traditionally Aberdeen Angus/Galloway were breeds of choice, but latterly certain strains of Hereford have proved suitable for crossing in most dairy breeds.

The recent introduction of the French beef breed Salers would appear to be another alternative bull for easy calvings. A calf which has been subjected to a severe calving is often too weak and shocked to suckle well; also the calf's swollen head and tongue are a restriction to easy feeding.

Cattle must calve in a suitable box (if inside) where the newborn calf may soon dry out, stay warm and avoid the filth and dirt which are characteristic of many cattle yards. Immediately following birth, provided checks on cow and calf are satisfactory, it is advisable that cow and calf bond together, i.e. get to know and accept each other.

It is advisable at this stage to dress the navel in iodine or with an antibiotic aerosol.

Lively calves will soon be up and about and take their first suck of colostrum. Note that in the absence of an adequate supply of colostrum from the cow, stored colostrum may be fed.

Calf Enteritis and Pneumonia

The main causes of unthriftiness and death in calves are enteritis and pneumonia. Everyone is familiar with the scouring calf and the coughing calf.

There are many feeding faults leading to scour, the most obvious being:

- inaccurate mixing of the milk replacer
- mixing and feeding at too high a temperature
- variable intervals between feeds
- overfeeding
- feeding contaminated milk

It is much easier for infective organisms to challenge a calf which has indigestion or an inflamed bowel.

As our knowledge progresses, it is obvious that there are many more causes than the simple answers of E. coli, salmonella and virus

Table 5.2   Calf enteritis

Disease organism	Frequency in cases diagnosed	Area of bowel affected	Age affected	Comments on husbandry system
Rotavirus	Over 90%	First part of small intestine	2–20 days	Dairy herd Beef suckler herd
Cryptosporidia	Over 50%	All small intestine	2–20 days	Single and multiple suckler beef
K99+ E. coli	Less than 5%	Lower part small intestine	Under 7 days	Dairy herd and beef suckler herd
Coronavirus	Over 50%	Any part of small intestine	2–25 days	Difficult to identify and diagnose
Calici-like viruses	Over 25%		2–20 days	
BVD	Approx. 5%	Mouth and entire bowel	10 days–2 years	Any system
Salmonella	Varies	Stomach and small intestine	1 day–8 weeks	Bought-in calves

Many outbreaks of enteritis may have two or more organisms involved. In a recent survey the most frequent of such combinations were:

Rotavirus and Coronavirus
Rotavirus and Cryptosporidia
Rotavirus and K99+ E. coli

In contrast to this, salmonella often acted as a single cause. Rotavirus and cryptosporidia were often found in the faeces of healthy calves.

pneumonia; for that reason the causes and diagnosis will be discussed in greater detail in this chapter.

All too often the calf rearer is presented with the situation of a calf with diarrhoea or one which is blowing. Has the calf just eaten too much, or has it increased respirations from the humidity, or heater fumes in its environment? Sometimes these suppositions are correct, but also these signs are the first ones of disease processes.

First, a careful examination of the calf and its companions is necessary. Second come isolation and treatment of any diseased calves. Third, the environment and feeding must be re-checked in order to avoid further problems.

From these two conditions morbidity varies from 10 to 40 per cent and mortality is approximately 5 per cent in a typical calf unit. The period that the calf is affected is commonly 10 to 15 days (range 0 to 40 days).

Rotavirus

This is the commonest causal organism of calf scour and is mostly associated with suckler units. Adult cattle and healthy calves can carry the organism, and it is often poor husbandry or the added presence of other causes of enteritis which causes disease. There is a thin, profuse yellow scour, and calves are often depressed and weak.

No drugs are directly active against the virus, but colostrum has a highly beneficial effect both in the bowel contents and when absorbed into the calf's bloodstream.

K99+ E. coli

Escherichia coli (better known as E. coli) is a bacterium; there are many different types and strains of the organism. In the past most cases of calf diarrhoea were diagnosed as E. coli because the bacteria was found in the faeces. It is now considered that only certain strains are responsible for disease, and foremost among these is strain K99+, which causes baby calf scour and sometimes septicaemia (the germ invades the calf's whole system via the bloodstream), often with fatal results.

Affected calves have loose yellow scouring; they are weak and rapidly become dehydrated and shocked. Quite often there is no rise in temperature, sometimes a subnormal reading.

Many outbreaks are associated with rotavirus. In order to minimise infections, strict hygiene and good husbandry are required for the newborn calf.

E. coli are usually susceptible to a suitably chosen antibiotic; colostrum is the main means of preventing this infection.

Cryptosporidia

Cryptosporidia, which are a form of coccidial parasite, can affect all species of animal (including man) and are distributed worldwide.

Infection is by oocysts from faeces, or by free living forms in the environment. Infection can remain for up to six months in buildings, and elimination by disinfection is difficult, the greatest success being by steam cleaning using 10 per cent formalin or not less than 2.5 per cent ammonia.

Affected animals are often depressed and have a poor appetite and a temperature up to 40°C. Faeces are pasty and yellow and later contain mucous.

There is no specific treatment against cryptosporidia. Colostrum has no specific effect on the disease other than to control other organisms such as rotavirus and K99+ E. coli.

Other forms of coccidia which may occur in calves are Eimeria bovis and Eimeria zuerni at around three weeks to three months of age.

Salmonella

There are many types and strains of salmonella bacteria, the most common associated with cattle being S. dublin and S. typhimurium.

Salmonella may affect all species including man, and when diagnosed, constitute a notifiable disease. Infection may be via milk, saliva or an environment contaminated from the faeces of vermin, adult cattle or other calves.

The duration of infection in buildings may be six months. The effect on animals can vary from scouring to an acute septicaemia and death. Calves often have a profuse watery scour, which may also contain blood and mucous. Their temperature can rise to 40.5°C.

In recent years a strain of salmonella known as salmonella 204C has increased in its resistance to antibiotic treatment, and has also become more ferocious in its attack on calves. The result has been a rapid spread of this disease throughout the country, many typical outbreaks being associated with calf rearing units which rely on calves from markets and dealers. Infection in a group can start after a few days, reach a peak after two to three weeks and start subsiding by the sixth week. Most salmonellae are susceptible to a suitably chosen antibiotic apart from the one quoted above.

Do remember that certain disease organisms may affect people. The most obvious of these are cryptosporidia and salmonella, and farm workers should be advised of this danger.

Calici-like Virus and Coronavirus

Calici-like viruses and coronavirus are more difficult to diagnose and their true significance in outbreaks of diarrhoea has yet to be fully investigated. They tend to occur more in the south of England.

Bovine Viral Diarrhoea

This virus is not usually associated with calfhood bowel problems and rarely may be the cause of a severe and chronic diarrhoea, with or without pneumonia, which damages and ulcerates the whole of the bowel from mouth to rectum and leads to death, or destruction on humane grounds.

It has been seen in calves from as young as ten days, but is usually found from two months of age or even later. It is mentioned because of the danger the virus may cause, especially if contracted by pregnant cattle. Cows infected during the first 100 days of pregnancy may give rise to unthrifty calves which constantly excrete the virus, and may themselves later die. Cows may also abort or have deformed calves.

For more information ask your veterinary surgeon about the Bovine Viral Diarrhoea/mucosal disease complex. Antibiotics may be used to control secondary infections.

Calf Pneumonia

Whereas many calf scours occur in the first week or two after birth, pneumonias are most prevalent one to three weeks after purchase or being rehoused. Cattle of different age groups should not be housed under the same roof because of the risk of cross-infection with respiratory disease.

Table 5.3   Infectious calf pneumonia

Disease organism	Area of respiratory tract	Age	Husbandry system
Parainfluenza type 3 (PI3)	Upper	2–4 months	Badly ventilated calf houses
Infectious Bovine Rhinitis (IBR)	Upper; also eyes and nose	All ages	Introduction of new animals to herd
Respiratory Syncytial Virus (RSV)	Lower; front area of lungs	Young calves mainly	Badly ventilated calf houses
Pasteurella pneumonia	Upper and lower	6 weeks, also adults	Transport and times of stress, rehousing and chills
Mycoplasmas	Lower	3–8 weeks	Often with other pneumonia organisms

There are several viral and bacterial causes of pneumonia, which may occur alone in a calf or combine to cause the infection. The basic essentials to minimise the risk of pneumonia are good ventilation, low humidity and a warm dry area for resting. Remember that too much restriction of the calf's movement in cold weather increases the chance of it being chilled.

Parainfluenza Type 3 (PI3)
This virus is a common cause of pneumonia and causes high temperatures, coughing and respiratory distress.

Whenever a percentage of a group of calves is affected, it is wise to treat the whole group by injectable antibiotics and supporting drugs (see later in this chapter). The viruses are not killed by antibiotics, but these drugs lessen the chance of a further bacterial pneumonia and abscess formation in the lungs. There are vaccines against PI3.

Affected calves are usually 2–4 months of age.

Respiratory Syncytial Virus (RSV)
This virus, which is the most common cause of viral pneumonia in calves, is particularly severe. It attacks mainly young calves and can often be fatal after a short illness. There is a vaccine against this infection.

Affected calves do not always cough, but they run high temperatures and may blow rapidly. Most cases occur over the winter months and are worse in times of bad weather.

Infectious Bovine Rhinitis (IBR)
This virus affects cattle of all ages and is thus a danger to the suckler as well as dairy cow adults. It can cause brain damage in calves, but one of the greatest risks is abortion in adult cows. Infections, which may be fatal, are of the upper respiratory tract, and hence there is a profuse discharge from the nose and infection under the eye lids. Mild cases have been mistaken for New Forest eye infection.

There is a good vaccine available.

Pasteurella Pneumonia
Pasteurellae are bacteria which cause pneumonias in several species of animals. There are various strains:

• Pasteurella haemolytica type A1
• Pasteurella multocida
• Pasteurella septica

These bacteria may cause pneumonia alone, but are often secondary invaders following infection with one of the viruses. To a lesser extent other bacteria may act in this way, e.g.

- Haemophilus species
- Mycoplasma bovis
- Bordatella
- Streptococci
- Corynebacteria and staphylococci

Affected calves cough and run high temperatures. Pasteurella may invade both upper and lower areas of the respiratory system, and often like Infectious Bovine Rhinitis cause infection on and around the eyes. In adult cattle, transport may precipitate an outbreak of pasteurella pneumonia—this is known as transit fever.

There are vaccines available against these bacteria, and they are also susceptible to some antibiotics.

Bovine Viral Diarrhoea
This is not a primary respiratory virus, but may cause pneumonia as well as scour in a few susceptible calves.

Inhalation Pneumonia
This is caused by 'force feeding' calves with a fluid feed or milk when they will not readily drink; it may also follow inefficient drenching when the medicine enters the calf's lungs instead of being swallowed. The severity of the condition depends on how much fluid and compound has travelled down the windpipe to the lungs. In all cases an antibiotic injection is advised to prevent a bacterial pneumonia arising.

Samples Required for Diagnosis

Rotavirus	Sample of faeces.
K99+ E. coli	Swab from faeces. Heart blood from dead calf.
Salmonella and cryptosporidia	Swab from faeces and from housing. Blood samples and faeces.
BVD	Blood samples. P.M. evidence of bowel.
PI3	Long swab from pharynx. Blood sample. P.M. evidence from lungs.
RSV	Swabs from eyes and nose. Blood samples.
Pasteurella	Swabs from eyes and nose. P.M. evidence from lungs.

Swabs from respiratory viruses are not merely wiped across the affected area; they should be rubbed on the nose or throat lining in order to pick up minute tissue cells which contain the virus. Note that a fresh sample of faeces (about 20 gms) is required to diagnose rotavirus or cryptosporidia. A dead calf will often on post-mortem provide all necessary samples for examination.

Blood samples will in the case of some diseases yield the living organism, e.g. the virus in BVD, but more often the blood is analysed for the antibodies (and their level) for a particular disease, and to complete this evidence a second (or paired) sample is often taken and analysed two weeks later from the same animal.

When dealing with diseased animals it must be remembered that some diseases are zoonoses, i.e. may infect susceptible people. Examples are salmonella, cryptosporidia and possibly RSV.

Treatment of Enteritis and Pneumonia

There have been major faults in the past regarding treatment of both enteritis and pneumonia in calves.

Too often calves have been abused with the indiscriminate use of antibiotics. It is only bacteria, E. coli, salmonella and similar organisms, that are killed by antibiotics and it is necessary to use the correct antibiotic for a particular bacteria.

In bacterial enteritis it is advisable to give the selected antibiotic by injection; only in pressing circumstances should the drug be given orally as well. Other drugs given orally are ones which have an absorbent capacity, e.g. kaolin, pectin and charcoal, and for anti-secretory effect, atropine and morphine.

One of the most significant changes in calves with severe or persistent scouring is dehydration. This means that a percentage of the body fluids is lost, the severity of the effect depending on the degree of fluid loss. A calf with a 5 per cent loss shows weakness, loss of weight, depression and a slowness in all its activities. Approaching 10 per cent loss of fluid, a calf is in a severe state of shock, often lying in a collapsed state and close to death in an irreversible condition.

It is known that pinching the eyelid of an affected calf shows the degree of dehydration by the length of time it takes the skin to flatten out again. Any obvious distortion of the skin for longer than two seconds indicates that the calf requires fluid. To rehydrate the calf it is advisable to stop feeding with milk or water and give it by mouth a glycerine and salt solution (electrolyte solution). If the calf will not drink this, it should be administered with an oesophageal tube. Feed approximately three litres morning and evening for two days; for the next two days feed half electrolyte solution and half milk.

A typical formula for 40 grams of such an electrolyte/glycerine preparation is:

20 gm glycerine
10 gm sodium chloride
 7 gm monopotassium phosphate
 2 gm calcium gluconate
0.5 gm magnesium sulphate

Some preparations also contain glucose.

Unfortunately calves may be so severely affected that they require further fluid, and this is best given by intravenous injection into the jugular vein. Fluid given under the skin to a severely shocked calf is not very effective because in shock the blood vessels are restricted and will not absorb much fluid from the area.

All scouring calves should receive oral electrolyte therapy regardless of the cause of enteritis.

Probiotics

It would be fair to say that probiotics for calves currently have a high profile. They have 'arrived' at a time when there is much concern amongst farmers because of health problems in all intensive livestock units; and this in spite of heavy use of very expensive conventional drugs—the antibiotics.

This situation is most vividly illustrated by the high mortality rate of calves infected with salmonella typhimurium Type 204C, which is totally resistant to all antibiotics available, including chloramphenicol and apramycin. With this situation all those concerned with animal health and welfare are stressing the need to go back to the basis of calf health—in particular the intake of vitally important colostrum, together with good hygiene and housing and anything which will reduce stress.

Probiotics, literally meaning 'for life', are being promoted as:

1. Natural
2. Anti-stress
3. Relatively cheap and available without prescription

They have an attraction to some calf rearers, and their use fits the general public and media pressure on farmers to move to no additives in human food or for animals producing this food.

What Is a Probiotic?
Probiotics are generally defined as live preparations of either one or a

mixed strain of various lactic acid bacteria, usually lactobacilli or streptococci.

In order to bring positive benefit, they must be viable in large numbers right up to the moment when they arrive in the gut of the calf. For these reasons some products are prepared by freeze drying. The numbers of individual bacteria are indeed very large and are expressed as 10 to the power of 6 (10^6), or 10 million, up to 10 billion per gram (10^9). It is necessary to emphasise that the digestive tract of the healthy calf is colonised naturally by similar large numbers of these same organisms, which are attached to the mucous membrane or free in the lumen of the gut.

It is important to realise that there is a tremendous difference in the physical and chemical environment of the various parts of the digestive tract, particularly in terms of pH and level of enzyme production. For example, the pH of the abomasum (true stomach) prior to milk feeding is around 2, i.e. very acid. The change through the pyloric sphincter (valve) into the small intestine is dramatic, rising to ph 8. This change is in addition to the presence of the powerful fat-splitting enzymes and emulsifiers in bile, other protein digesters and lactase to digest lactose.

To have any beneficial effect in the calf, the live organisms in a probiotic preparation administered within a milk substitute need not only to survive the conditions of the abomasum and enter the lower gut, but to multiply and dominate the existing flora.

For the probiotic to be successful, the additional complication in the calf lies in its developing rumen, which is nothing less than a microbial fermentation vat. Again, the concentration of bacilli is great (up to 10^{13} of lactobacilli and streptococci) and in total volume terms expanding to several litres of rumen contents very quickly. The organisms herein also include types which are different from those involved in sugar fermentation lower down the gut. They have to deal with a wide range of carbohydrates, including digestible fibre. For this role the larger protozoa, as well as bacteria, are needed.

Much is known about the conditions which are favourable for good rumen fermentation and development of the rumen papillae. In general the diet needs to include materials which are not too readily fermentable and as a result maintain the pH nearer to 6 than 5. Probiotics included in any feed would have no chance of passing unchanged through the rumen to the lower gut.

It can be seen that for a probiotic to meet all the varying requirements of the conditions in the digestive tract is expecting much.

How Do Probiotics Work?
The organisms in probiotics are all fermenters of lactose. When lactose

passes from the abomasum into the small intestine, it is split by the enzyme lactase into two simple sugar molecules, prior to absorption into the blood stream. The two processes are thus running in parallel.

The lactic fermentation probably dominates the processes and in particular keeps the undesirable E. coli organisms in check, for it is these bacteria which give rise to scouring in the calf. The mechanisms involved may include lowering the pH, competition for nutrients, blocking of binding sites and anti E. Coli metabolites.

Scouring is most likely to occur under the adverse conditions which arise from poor-quality milk powders, irregular feeding and, in particular, stress conditions caused by the movement of calves through markets. This involves repeated (mis)handlings, missed feeds and physical extremes of heat or cold.

It is logical that probiotics are most likely to be of benefit following these adverse conditions. When calves have become ill and dehydrated, they are often treated with antibiotics by mouth and/or by injection. Another potential role for probiotics is to quickly re-establish the correct flora population. In order to do this they must supply an inoculum of the organisms missing, be compatible with the animal species and its relevant intestinal environment, and possibly be resistant to the antibiotic in question.

Recommendations by the manufacturers of probiotics suggest the need for very large doses of organisms, either in daily amounts of, say, 1 million per gram of feed or as several billion in one dose at critical times. These numbers compare with bacterial populations in whole milk ranging from 200,000 to 1 million per ml, or in milk powders of around 50,000 per gram. Most of these organisms are lactose fermenters just as in the probiotics.

Assessing the Benefit of Probiotics
This is the difficult part. First of all there are a number of products being sold, some including a single species and others with a 'cocktail' of organisms. Also milk powders are being marketed which just claim 'Probiotics added'.

The farmer is probably looking for increased productivity. This can be measured in terms of better health (probably the most desirable effect), but also by increased growth rate. Part of the improvements to health could arise, not just from the effect of the production of lactic acid and the suppression of undesirable organisms, but from a process known as immunomodulation.

This simply means that the normal protective immune processes, including the proliferation of lymphocytes, may be stimulated by the lactobacilli. There are other microbiological substances, and also various drugs like the worming preparation levamisole, vitamin E or zinc, which are known to have a similar effect.

Much of the early work on probiotics has been done with pigs and there is virtually no independent long-term trial work with calves in this country. Several trials have been conducted by BOCMS at Barhill Development Farm and on a field trial farm, and no statistically significant improvements in performance or health were found. Indeed in one trial, higher rates of probiotic dosage were found to depress dry feed intake. In contrast there are some claims being made publicly endorsing the general inclusion of probiotics in calf diets which, from our long-term knowledge of the problems of finding valid results working with young calves, would seem extravagant. However, many rearers are convinced they are of benefit and retailers of the products are enthusiastic too. It is easy to be critical of individual trials. For probiotics to be effective it can be postulated that calves must be under stress or below average health for a benefit to be seen. For these reasons the only way to hope to demonstrate probiotic efficacy is with well-controlled trials which record results including bacteriology over a period of time. This sort of work is expensive to undertake and can only examine one or two products. Studies could however be made on a factorial basis in other trial work.

One further complication is that part of the ethos of probiotics is that they 'pre-empt' the onset of disease, though none would claim that calves can be protected against virulent organisms like the salmonellas. Additionally, the major problem on most calf units is with pneumonia, i.e. not in the digestive tract at all, though the two exacerbate each other.

For those committed to natural prevention of disease, probiotics would be preferred to the inclusion of feed antibiotics or growth promoters. However, all commercial calf feeds, both milk replacers and dry feeds, now contain a growth promoter (e.g. avoparcin, virginiamycin or nitrovin), and these would therefore need to be excluded from or only form part of the trial as one of the factors in any comparison.

In summary:

1. Probiotics are preparations of live, naturally occurring bacteria, usually lactobacilli or streptococci, which ferment lactose.
2. Currently the calf rearer has to take on trust the data on numbers of live organisms and their shelf life.
3. In general, it can be assumed that these organisms would not withstand the temperatures reached when dry feeds are pelleted.
4. The provision of these extra bacteria into the *rumen* would seem to be unnecessary.

5. They are best directed into the true stomach and small intestine, especially at critical times, and following antibiotic treatment. This will normally be during the pre-weaning period.
6. Probiotics are not expected to deal with severe infections of the gut or other parts of the body.
7. It will probably be a number of years before independent definitive trial results are available. Those so far carried out are inconclusive and some show a depression in any dry feed intake.

Antibiotics

Most viral pneumonias are complicated by the fact that bacteria attack the inflamed or diseased respiratory tract, hence the need for adequate antibiotic therapy by injection.

In the past, sole reliance was placed on such drugs, but recently great progress has been made both in the speedy and full recovery of affected calves and in saving greater numbers of calves by the use of other injectable compounds with selected antibiotics. These include drugs such as:

- Flunixin meglumine, which reduces the inflammation, pain and temperature associated with an infective pneumonia.

Table 5.4 Vaccines available, time and route of administration and efficiency

Vaccine	Time	Route	Efficiency (Personal Opinions)
Rotavirus	Late pregnancy	Cow injection	Good
K99+ E. coli	Late pregnancy	Cow injection	Good
Salmonella (certain strains only)	Young calf	Calf injection	Fair
RSV	Young calf	Nasal drops Calf injection	Good Poor
IBR	Young calf	Nasal drops	Good
PI3	Young calf	Nasal drops Injection	Good Poor
BVD	Young calf	Calf and adult injection	Poor
Pasteurella	Young and older calves	Calf injection	Good
Mycoplasma	Young calf	Calf injection	Insufficient information

- Bromhexine hydrochloride (Bisolvon), which breaks up the accumulated mucous in the bronchial tubes of the lungs, allowing freer air flow and better oxygen exchange.
- Clenbuterol hydrochloride, which dilutes the bronchial tubes, again allowing better air flow and less consolidation of lung tissue.

Bromhexine and Clenbuterol may be included in feed. Terramycin is now combined with some of these compounds for easier administration by injection.

Although most enteric and pneumonic bacteria are sensitive to Chloramphenicol, medical and veterinary authorities advise that the drug should be used only for human treatment, and it is hoped to avoid further increases in bacterial resistance by indiscriminate use in animals.

Most pneumonias respond best to the drug Trimethoprin and the antibiotics terramycin and penicillin.

Remember that in order to avoid relapses in pneumonia, treatment with antibiotics should be of sufficient high dosage and duration of treatment to prevent recurring bacterial infection.

Use of Veterinary Medicines

The Animals, Meat and Meat Products (Examination for Residues of Maximum Residue Limits) Regulations 1991 states that it is an offence to use unauthorised medicines or to sell an animal which has drug residues in it. Medicines must be secured in a locked compartment, and their use recorded within 72 hours. NOAH (National Office of Animal Health Limited) and AHDA (Animal Health Distributors Association) have printed a record book which farmers may use for this purpose.

Prevention of Enteric and Pneumonic Infections

Let it be stated again that good housing, good husbandry and good food are essential for reasonable results in calf rearing. Further measures may be necessary to obtain protection against some infective agents. Vaccines are available against many of the infective agents already discussed, some against one organism and others against several. The benefits from vaccines vary because some give an enormous boost to immunity against disease and others hardly seem to alter immunity against another organism.

Some vaccines against enteritis in young calves are best administered to the cow, in order to boost antibodies in the colostrum and thus

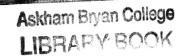

protect the calf. Most pneumonia vaccines are given when the calf is a few weeks old to protect it during the growing period. However, for immediate protection a sero-vaccine can be given, boosted by a second dose of just the vaccine two weeks later.

Note that enteric vaccines given to cows are of no use unless the calf receives adequate colostrum during the first three or four days. Calves require at least two litres of colostrum in the first twelve hours. Vaccines given to calves in early life often need a second boost six or eight weeks later; however, if the first administration is at eight weeks, a single dose may suffice. My own personal opinion is that wherever there is a great risk of young calves developing pneumonia, a vaccine which gives protection against several types of pasteurella is a wise insurance.

Many vaccines are available against more than one organism, and some combinations are shown below:

Rotavirus and K99+ E. coli
Salmonella typhimurium and Salmonella dublin
Parainfluenza and Infectious Bovine Rhinitis viruses
Mycoplasma, Respiratory Syncytial Virus and Parainfluenza virus

Remember that calves may need vaccines to protect them against diseases found in the adult herd or on the farm pastures. Vaccines for this purpose are available against *Leptospira hardjo*, blackleg, tetanus, Infectious Bovine Rhinitis and oral lungworm vaccine.

One risk is that disease may strike several calves before the vaccine has had time to be useful; this is particularly true of Respiratory Syncytial Virus, where the young may be affected with fatal results.

Blanket cover with all vaccines is a waste of money, as is just guessing at which vaccine may be the best to employ—the guess can easily be wrong.

Resuscitation

The most serious physiologically damaging factor caused by calf enteritis is the dehydration/shock syndrome that may occur. Shock is a complex result of many diseases or injuries, but is most often associated with the severe dehydration that can follow haemorrhage, scour or inflammation of the calf's alimentary tract (bowel). The loss or redistribution of large volumes of body fluids due to the passing of large quantities of fluid faeces often leaves the calf in a state of collapse.

Apart from the physical assessment of dehydration previously described, e.g. pinching of eyelids, other tests may be applied. A small blood sample may be taken to determine the haematocrit (the blood's relative volumes of cells to fluid). This is an indication of the degree of dehydration. The calf's blood may be analysed using a 'blood gas machine' to determine the degree to which it varies from the correct range of pH (acidity/alkalinity). Following these analyses, fluid of the correct makeup may be given intravenously in order to restore the correct body fluid volume, minerals content and pH. As much as 18 litres over 48 hours may be given. All such veterinary attention needs to be given promptly and carefully as soon as signs of problems have been detected in the calf. Delays cost further losses and expense.

Nursing

Due to the isolation of farms, the cost of professional advice, and sheer necessity in an emergency, the farmer and his staff should be able to cope with sudden illness, and aid weak or sickly calves. The procedures they adopt should be adequate and the staff should be confident in applying them.

The best example is the bloated calf just referred to. Sick calves often need isolating from a bunch, into temporary accommodation, which can be made up from hurdles and bales. This will keep the calves warm, isolate them as a source of infection and give them plenty of undisturbed rest. If a heat lamp is provided care in siting must be observed to avoid burning or electrocution.

The calf should not be allowed to lie in one position for too long or congestion of the lungs will occur. It should be turned from time to time and when possible propped up in the correct sitting position with the front legs tucked underneath. Dehydration is often the cause of calf death, and where a calf will not drink, a proprietary plastic oesophageal tube is an excellent way of administering electrolyte (specific salt mixture) solutions to maintain the body fluids. Calves which have collapsed may be given suitable solutions intravenously or under the skin to reverse a condition of shock.

Necessary injections or tablets should be administered correctly by sterile equipment, and the doses and length of treatment advised by the veterinary surgeon adhered to. A record in the 'sick book' of all data relating to ill calves is to be strongly recommended, especially on large units and where staff change at weekends.

Care and patience in treating sick calves is an obvious virtue, combined with commonsense and keen observation. Good communication between rearers and their professional advisers will undoubtedly lead to successful calf rearing.

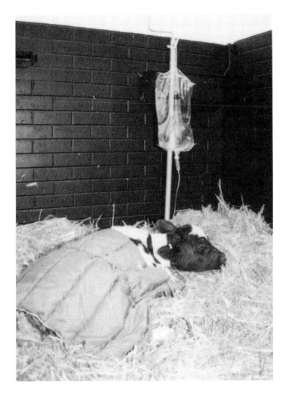

Plate 30   A dehydrated calf
being resuscitated by
administration of intravenous
fluids

WELFARE

Basic interrelated factors are of importance under this heading. First, the fundamental welfare requirements of calves must receive satisfactory attention. Second, there are laws and codes of practice regarding farm animal welfare which farmers, haulage contractors and all who are associated with responsibility for calves in their charge will abide by.

It is a self-obvious fact that calves reared in a system which attends satisfactorily to their welfare stand a good chance of thriving within that environment and that these calves are most likely to prove of economic benefit to the farmer.

Professor John Webster in his book *Animal Welfare* describes how the basic principles stated by the Brambell committee (1965) and recently revised by the Farm Animal Welfare Council (1993) may be outlined, as shown:

1. Freedom from thirst, hunger and malnutrition—by ready access to fresh water and a diet to maintain full health and vigour.
2. Freedom from discomfort—by providing a suitable environment including shelter and a comfortable resting place.
3. Freedom from pain, injury and disease—by prevention or rapid diagnosis and treatment.
4. Freedom to express normal behaviour—by providing sufficient space, proper facilities and company of the animal's own kind.
5. Freedom from fear and distress—by ensuring conditions which avoid mental suffering.*

I would include in welfare the necessity for an overall plan for the calf rearing system that prevails on each particular farm. This plan should outline correct housing or environment which satisfies the five points outlined above.

Preventative measures must be adopted regarding disease risks either on the farm (e.g. presence of IBR or BVD on the premises) or likely to occur there because of a specific husbandry system used (e.g. buying in calves). Preventive measures may dictate the size of calf groups for rearing, the use of vaccines, or vigilance regarding a specific symptom of a disease most likely to occur.

Calf welfare is a responsibility and obligation to which farmers must respond vigorously. The Welfare of Animals during Transport Order 1994 (WATO) now applies. This order states that no unnecessary suffering may be caused as a result of transport, and in particular an animal may not be transported if it is:

1. new-born
2. diseased
3. infirm
4. ill
5. injured
6. fatigued
7. has given birth within the last 48 hours
8. is likely to give birth during transport
9. or is unfit for any other reason

All cattle transported must be fed, watered and rested before the journey and not more than 15 hours after the start of the transport. No Journey Plan or Transport Certificate is required for journeys of less than 30 miles to and from agricultural land, if the vehicle is owned by the landowner and the internal length of the vehicle is less

---

* *Animal Welfare: a cool eye towards Eden* (Oxford: Blackwell Science, 1994), p. 11. Reprinted by permission of the publisher.

than 3.1 metres. Otherwise journeys of less than 15 hours require a Journey Plan shown on an Animal Transport Certificate (which must be retained for six months).

For journeys over 15 hours, a Journey Plan must be carried by the transport and give details of resting, feeding and watering (this must be countersigned by the consignor at the end of the journey and a copy sent to the DVO). This order gives further requirements, duties and vehicle construction regarding transport of animals. The above came into force January 1995.

ZOONOSES

This refers to diseases which are present in the animal population, and can also cause disease problems in man.

Salmonella is an obvious example. Many of the Salmonella types can cause food poisoning of various degrees of severity, e.g. *Salmonella enteriditis*, *typhimurium*, *dublin* etc.

Severe outbreaks of enteritis in calves may be caused by Salmonella type 204C. This organism is particularly virulent and resistant to many antibiotics currently in use. This is one of the reasons why it is now an offence to use chloramphenicol in food-producing animals. The drug is reserved for the treatment of the human population.

Another cause of disease in persons is the verotoxigenic *Escherichia coli* (VEC), which is responsible for haemolytic uraemia syndrome and haemorrhagic colitis, both of which can be life threatening.

Cryptosporidia is a third cause of calf scour which can and does affect people.

All cases of calf scour/enteritis should be treated as potential human disease and strict hygiene precautions taken by those living or working on the farm to avoid cross-infection.

Ringworm can cause human infection. Prompt and effective treatment and strict hygiene are necessary.

Leptospirosis (which can give severe flu symptoms) is a risk to human health and an organism readily transmitted through untreated milk. *Leptospira hardjo*, which causes abortion in cattle, is the most likely organism of this type to affect farm workers.

CALF SURGERY

Castration and Debudding

According to the Protection of Animals (Anaesthetic) Acts of 1954 and 1964, an anaesthetic must be used for:

1. The castration of a bull by means of a device that constricts the flow of blood to the scrotum, unless the device is applied within the first week of life.
2. The castration of a bull by any means once it has reached the age of three months.
3. The dehorning of adult cattle.
4. The disbudding of calves, except by chemical cauterisation within the first week of life.

Castration and debudding of calves should be done when calves are four to six weeks of age; only a veterinary surgeon may castrate a bull which is over twelve months of age. At the time calves should be healthy, active and free from clinical signs of disease or injury and it is also wise to make sure these surgical interferences do not coincide with transportation or rehousing.

Either employ a veterinary surgeon or ensure that the person who does the debudding and castration is adequately instructed in the method to employ and that the welfare requirements (need for anaesthetics) and procedures are carried out carefully.

Methods
My personal opinion is that caustic applications to the horn buds can cause suffering and are not always effective. It is much wiser to use an electric or Calor gas heated iron for horns and to do the job effectively. Recently electric irons which blow very hot air have proved satisfactory.

Elastrator bands to the scrotum may also cause suffering, can lead to chronic infection (including tetanus) and if great care is not taken will leave a testicle (under the skin of the body wall) even though the scrotum has separated. Castration by knife and castration by Burdizzo (bloodless crushing of the vessels to the testicle) are equally effective and satisfactory.

Teats

Removal of Supernumerary Teats from Heifers
Heifers intended for future breeding or milking should be examined when they reach one month of age in order to determine whether the teats and their arrangement are consistent with normal suckling on four quarters when they mature. For this purpose, it is necessary to check on the placing of teats and decide whether there are any extra (or supernumerary) ones.

Method
Select the teat or teats that are to be disposed of, wash the area and

inject a small blob of local anaesthetic at its base. With a small Burdizzo castrator pinch the teat close to its base. With clean, sharp scissors cut the teat away along the crush line left by the castrator. Apply a little antiseptic powder or antibiotic aerosol.

Navels

All calves' navels should be checked and dressed with a strong iodine solution as soon as possible after birth. One of the main reasons for interfering with calves, other than obvious cuts and injuries, is a problem arising from a bleeding or diseased navel.

At birth the umbilical cord vessels may be bleeding. Often this is only for a short period of time while the vessels are contracting, but the haemorrhage may be serious. In these circumstances it is often necessary for the person present to tie the vessels—preferably close to the body. This interference should be done as hygienically as possible, using for example a length of bandage. The extra navel tissue should be cut away two inches below the ligature and the whole dipped in a strong iodine solution. The ligature should not be left on for more than twelve hours, and the area should be rechecked at frequent intervals after it has been removed.

Unfortunately, the body wall may sometimes be open in the navel region and a prolapse of intestine occur through this hole. In these circumstances protect the bowel from becoming dirty (tie a clean sheet or towel around the calf, covering the prolapsed bowel) and seek immediate veterinary help.

Ruptures and abscesses may occur in the navel region, and expert help is sometimes needed to diagnose the cause of the navel swelling.

IDENTIFICATION

The Bovine Animals (Records, Identification and Movement) Order 1995 (BARIMO) is now in effect.

New ear tags were compulsory from 1 April 1995 and components to the front and back of the ear must give the following details: country of origin, herd mark and individual animal number. Tags are often plastic.

Dairy calves must be identified with the MAFF approved tags within 36 hours of birth, beef calves within 30 days of birth or sooner if their movement needs to be recorded. All lost tags must be replaced within 36 hours and the new number recorded.

Records should be kept of all births, deaths and movements of bovine animals from the day of birth. The format of a typical Move-

Plate 31   New approved tags

ment Book is illustrated on pages 160–61. This shows the details which must be entered. Note that these records must be kept for ten years.

REGISTRATION OF PREMISES

All premises on which cattle are kept must be registered with the Ministry of Agriculture (Directive 92/102 EEC).

All keepers of cattle should have a holding number, herd mark and details of animals kept. Such registration is necessary so that cattle can be traced if necessary from birth and during their entire lives.

The Beef Special Premium and Suckler Cow Premiums are worked in conjunction with these details. Herds are assigned a Production Number once they have been registered for any of the above benefits.

IMPORTATIONS

Recent changes in EC legislation have brought about the rapid international movement of farm animals from farm to farm, provided veterinary certification is satisfactory at the farm of origin and there are no current prohibitions due to notifiable disease outbreaks which affect the movement of animals for that particular journey.

On Farm Movement

Name and address of person keeping the record: . . . .

To be filled in when applicable	To be filled in when applicable	To be filled in for each animal for each event				To be filled in when a calf is born	To be filled in when a tag is lost
Date of movement onto farm, or of birth, or of loss of ear tag	Date of movement off farm, or of death	Ear tag number	Breed	Sex		Dam's approved identification mark	Replacement tag number

The local Divisional Veterinary Office (DVO) of MAFF will be informed at his office by the veterinary officers in the country of origin of the likely importation. The farmer who is importing the animals must inform the DVO of the time the imported cattle arrive. All cattle must be treated for warble fly within 24 hours of their arrival. The importation must be recorded.

It must be remembered that imported cattle, although certified satisfactory to travel, may carry diseases which will put your own livestock at risk. It is advisable to isolate imported cattle and check with your veterinary surgeon about possible disease risks.

FURTHER INFORMATION

The Ministry of Agriculture, Fisheries and Food and the Scottish and Welsh Agriculture Department have issued leaflet no. 701, *Cattle*, giving codes of recommendations for the welfare of livestock, and this should be studied.

Other Ministry booklets and leaflets which are well worth reading are:

The Cleansing and Disinfection of Calf Houses L. 645
MAFF Prevention of Calf Diseases, leaflets 517 and 836
The Animal Medicine Record Book, which may be obtained from
NOAH, 12 Whitehall, London SW1A 2DY, or from Animal Health

and Breeding Record

. . . .

To be filled in when an animal is moved onto or off holding				*General Comments*
Age or date of birth (if known)	Premises from which moved and the name and address of person from whom delivery was taken	Premises to which moved, and the name and address of person taking delivery	Number of animals moved	Cause and Place of Death can be recorded here

Distributors Association, Gable Court, Parsons Hill, Hollesley, Woodbridge IP1 3RB, phone (01394) 410444, fax (01394) 410455. It is useful to keep details of drugs dispensed.

Request details from your local divisional veterinary office regarding the legislation which applies for dealing with farm casualties.

APPENDICES

1. Norms for Milk Replacer and Feed Intake and Liveweight Gains

THE FOLLOWING four graphs illustrate data collected from a large number of purchased Friesian bull calves reared at Barhill farm. They represent a reliable all-the-year-round norm against which individual batches can be judged. The number of weeks starts from the day of purchase at an estimated ten days of age.

In the first graph the large difference in milk replacer intake between the bucket and ad-lib systems is illustrated. The necessity for restriction

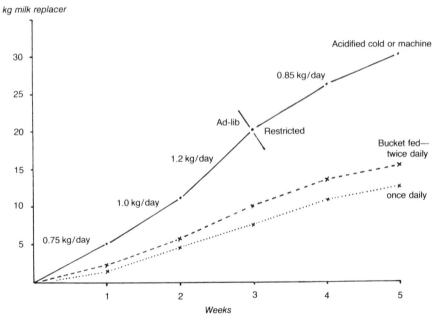

Graph 1   Cumulative intake of milk replacer 0–5 weeks

of intake on the ad-lib system at the three-week stage can be seen in order to achieve a smooth weaning at five weeks.

In graph 2 the S curve of the pattern of concentrate intake per day arises because of the slow build-up of dry feed appetite whilst the calves are on milk replacer. Once weaned at five weeks the calves' intake shows the most rapid increase per day, especially for those on the ad-lib system. Finally the increase in intake per day reaches a stage where it is governed solely by increasing liveweight. At twelve weeks or about 100 kg, dry matter appetite is about 3 per cent of liveweight, but by the time 500 kg liveweight is reached it will only be 2 per cent.

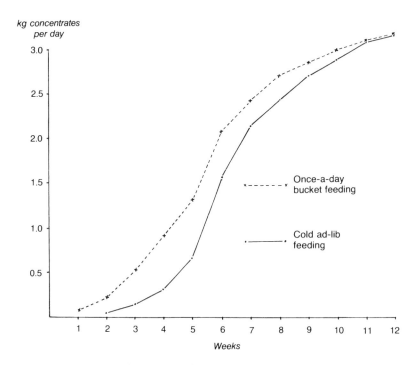

Graph 2   Daily intake of concentrate feed 0–12 weeks for Friesian bull calves

In graph 3 concentrate intake is presented from the same data in cumulative form. It shows that calves once weaned from either system eat dry feed at the same rate, though starting from a different base line. The graph enables quantities of concentrate required per calf to be calculated for both systems on the basis of free access throughout.

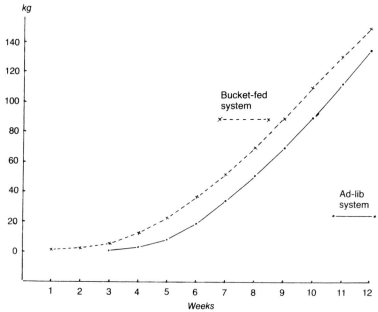

Graph 3   Cumulative concentrate feed intake 0–12 weeks for Friesian bull calves

In graph 4 the liveweight gains resulting from the concentrate intakes already given are shown for the two systems. The extra gain made by calves on the ad-lib system is all achieved by three weeks.

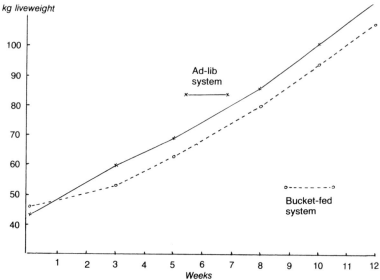

Graph 4   Liveweight gains 0–12 weeks for Friesian bull calves

2. Suggested Costings

Suggested costings for the four main rearing systems are given. When the different feed intakes and resulting performances are compared on a feed cost per kg of gain, it can be seen that the ad-lib systems are 28 per cent higher. This is because of the higher cost of gain made on milk replacer compared to dry feed. In money terms the bucket-fed calves will have a total cost around £24 less to take them to the same liveweight as the ad-lib fed calves, but over a week longer. This advantage has greatly increased over recent years as milk powders have escalated in price.

Suggested costing for rearing a Friesian bull calf on the once-a-day bucket system. 0–12 weeks. Liveweight gain 45–105 kg.

		£	£
Bought-in calf	say		140
12.5 kg skim based milk replacer	@ £1500	18.75	
154 kg early weaner dry feed	@ £210	32.34	
10 kg straw	@ £50	0.50	
Total feed cost			51.59
Feed cost per kg gain: 86p			
Variable costs			
Veterinary and hygiene		4.00	
Mortality 3% on £150		4.50	
Fuel and water		4.00	
Bedding 60 kg @ £40		2.40	
			14.90
Overhead costs			
Labour		15.00	
Depreciation, buildings and equipment		5.00	
Interest on capital @ 10% p.a.		5.00	
			25.00
Total Rearing Costs			91.49

Some of these costs would be ignored by an established rearer and treated as part of his profit.

Total costs per kg gain: 152p.
Liveweight gain per calf: 0.71 kg/day.

Use of a whey based milk replacer would give an approximate saving of £4 per calf.

When the best estimates of other costs are attributed to the four systems the costs per kg gain vary from 152p to 171p. Again it must be emphasised that individual rearers should put their own costs into something on the lines of these examples in order to pinpoint their strengths and weaknesses.

Other costs which may possibly be incurred at the customer's request are castration and vaccinations for I.B.R., salmonella or husk.

A profit figure would need to be added to the total costs in order to give a selling price.

Suggested costing for rearing a Friesian bull calf on the twice-a-day bucket system. 0–12 weeks. Liveweight gain 45–105 kg.

		£	£
Bought-in calf	say		140
15.6 kg skim based milk replacer	@ £1500	23.40	
150 kg early weaner dry feed	@ £210	31.50	
10 kg straw	@ £50	0.50	
Total feed cost			55.40
Feed cost per kg gain: 92p			
Variable costs			
Veterinary and hygiene		4.00	
Mortality 3% on £150		4.50	
Fuel and water		4.00	
Bedding 60 kg @ £40		2.40	
			14.90
Overhead costs			
Labour		16.00	
Depreciation, buildings and equipment		5.00	
Interest on capital @ 10% p.a.		5.00	
			26.00
TOTAL REARING COSTS			96.30
Total costs per kg gain: 160p			
Liveweight gain per calf: 0.71 kg/day			

Use of a whey based milk replacer would give an approximate saving of £5.50 per calf.

Suggested costing for rearing a Friesian bull calf on the ad-lib machine system or on the ad-lib cold system. Liveweight gain 45–114 kg.

			£	£
Bought-in calf	say			140
33 kg skim based milk replacer	@ £1500		49.50	
135 kg early weaner dry feed	@ £210		28.35	
10 kg straw	@ £40		0.50	

Total feed cost   78.35
Feed cost per kg gain: 114 p

Variable costs
Veterinary and hygiene	4.00	
Mortality 3% on £150	4.50	
Fuel and water	4.00	
Bedding 80 kg @ £40	3.20	
		15.70

Overhead costs
Labour	13.00	
Depreciation, buildings and equipment	6.00	
Interest on capital @ 10% p.a.	5.00	
		24.00

TOTAL REARING COSTS   118.05

Total costs per kg gain: 171p
Liveweight gain per calf: 0.82 kg/day

Use of a whey based milk replacer would give a considerable saving of approximately £11.50 per calf.

INDEX

168